The ARRL's General Q&A

Upgrade to a General Class Ham License

By:

Larry D. Wolfgang, WR1B

Production Staff:

David Pingree, N1NAS, Senior Technical Illustrator:
Technical Illustrations

Jayne Pratt Lovelace, Proofreader

Paul Lappen, Production Assistant: Layout

Sue Fagan, Graphic Design Supervisor: Cover Design

Michelle Bloom, WB1ENT, Production Supervisor: Layout

Published By:

ARRL *The national association for AMATEUR RADIO*
225 Main Street
Newington, CT 06111-1494

Copyright © 2002 by

The American Radio Relay League, Inc

Copyright secured under the Pan-American Convention

International Copyright Secured

This work is publication No. 278 of the Radio Amateur's Library, published by ARRL. All rights reserved. No part of this work may be reproduced in any form except by written permission of the publisher. All rights of translation are reserved.

Printed in USA

Quedan reservados todos los derechos

ISBN: 0-87259-858-6

First Edition
First Printing, 2002

This book may be used for General class license exams given through June 30, 2004. (This date assumes no FCC Rules changes to the licensing structure or privileges for the General class license that would force the Volunteer Examiner Coordinators' Question Pool Committee to modify the question pool.) *QST* and *ARRLWeb* (http://www.arrl.org) will have news about any Rules changes.

Contents

- iv   Foreword
- vi   How to Use This Book
- viii   ARRL Study Materials

- 1   Introduction
- 21   General Subelement 1
  Commission's Rules
- 53   General Subelement 2
  Operating Procedures
- 85   General Subelement 3
  Radio-Wave Propagation
- 105   General Subelement 4
  Amateur Radio Practices
- 133   General Subelement 5
  Electrical Principles
- 143   General Subelement 6
  Circuit Components
- 151   General Subelement 7
  Practical Circuits
- 157   General Subelement 8
  Signals and Emissions
- 169   General Subelement 9
  Antennas and Feed Lines
- 193   General Subelement 10
  RF Safety
- 223   About the ARRL
- 225   ARRL Membership Application

Foreword

Congratulations on making the decision to earn your General class Amateur Radio license! Enjoy the magic of radio at its best with worldwide radio communication on the Amateur HF bands. Earning a General class license will add to your enjoyment of Amateur Radio far beyond the effort required to learn the material to pass the exam. You will gain access to major portions of all our Amateur HF bands. You will be permitted to use a wide variety of operating modes on those bands.

Of course you will be anxious to operate single-sideband phone on popular bands such as 20 meters! Be sure also to sample the excitement of the digital modes like PSK31, where you can easily make contacts with stations you aren't sure you can *hear*, even using less than 5 watts of transmitter power (QRP). Try your hand at visual communications with slow-scan TV. Go ahead and make a Morse code contact, too. You'll find plenty of other operators eager to make your acquaintance using this mode, where skill is the common denominator. No matter what your favorite form of operating is, you are sure to find some like-minded hams on the HF bands. Try a contest, join a net, chase some DX. Have FUN!

For more than 85 years, ARRL has been the Radio Amateur's own organization. The Headquarters Staff in Newington, Connecticut is here to serve you, today's ham. If you are a member of ARRL, we thank you. If you are not currently a member, we have included an Invitation to Membership in the back of this book. You don't even need an Amateur Radio license to be a member. Nearly 170,000 ARRL members enjoy a wide range of membership benefits, including a monthly magazine, *QST*, which offers articles and columns for beginners as well as more experienced hams. We look forward to meeting your Amateur Radio needs for many years to come.

ARRL's General Q & A will help prepare you for your General class license exam. But you aren't just studying for a license exam, and we aren't satisfied just to help you pass that exam! We want you to enjoy every aspect of this avocation and to be prepared to serve your community. The ARRL offers many operating aids and technical material in the many books and supplies that make up our "Radio Amateur's Library." If you have a question about Amateur Radio, you'll find the answer in an ARRL publication. *ARRLWeb* (**http://www.arrl.org/**) is a great resource, with lots of information, including copies of operating aids, charts and information you can download for free. You can order any ARRL publication online at *ARRLWeb*, or you can contact our Publications Sales Office to request the latest publications catalog or to place an order. (You can reach us by phone — 888-277-5289 (toll free); by fax — 860-594-0303 or by e-mail — *pubsales@arrl.org* as well as at *ARRLWeb*.)

There is a Feedback Form at the back of this book. We'd like to hear from you — your comments and suggestions are important to us. Thanks, and good luck!

David Sumner, K1ZZ
Executive Vice President, ARRL
Newington, Connecticut
March 2002

How to Use
this book

To earn a General class Amateur Radio license, you must pass (or receive credit for) exam Elements 1, 2 and 3. This book is designed to help you prepare for and pass the Element 3 General class written exam. If you do not already have a Technician Amateur Radio license and valid credit for passing the Morse code exam you will need some additional study materials. In that case, see The ARRL Study Materials section on the next page and additional information in the Introduction chapter of this book.

The General class written exam requires that you know some basic electronics theory and Amateur Radio operating practices and procedures. You will learn more about the rules and regulations governing the Amateur Service, as contained in Part 97 of Title 47 of the Code of Federal Regulations — the Federal Communications Commission (FCC) Rules.

The Element 3 exam consists of 35 questions taken from the question pool in this book. A passing score is 74%, so you must answer 26 of the 35 questions correctly to pass. (Another way to look at this is that you can get as many as 9 questions wrong and still pass the test.)

The questions and multiple-choice answers in this book are printed exactly as they were written by the Volunteer Examiner Coordinators' Question Pool Committee, and exactly as they will appear on your exam. (Be careful, though. The letter positions of the answers may be scrambled, so you can't simply memorize an answer letter for each question.) In this book, the letter of the correct answer is printed in **boldface type** just before the explanation. If you want to study without knowing the correct answer right away, simply cover the explanation section with your hand or a piece of paper as you read down the page.

ARRL Study Materials

ARRL offers a variety of study materials to help ensure your success on exam day. For the Element 2 Technician written exam, ARRL's *Now You're Talking!* includes friendly, easy-to-understand theory and rules explanations. This book will also help you set up and operate your first Amateur Radio station. The entire Technician question pool is included. *The ARRL Technician Class Video Course* creates a classroom on your TV screen, with your own personal instructors to make sure you understand each topic. The Course Notes book includes the complete question pool. *ARRL's Tech Q & A* has brief explanations to go along with every question in the question pool, printed directly after each question. Perfect for a quick review before the exam or as a brief refresher for anyone already familiar with the material.

The ARRL General Class License Manual has detailed explanations of all the material covered by the Element 3 General class written exam. The complete General class question pool is included. *The ARRL General Video Course* presents the material for this exam using your TV-screen classroom. Full-screen graphics and your personal instructors lead you through the exam material. *ARRL's General Q & A* has brief explanations after each question to refresh your memory or review the material just before your exam.

ARRL's *Your Introduction to Morse Code* is offered as a set of two audio cassette tapes or two audio CDs. You will learn all the characters required for the Element 1, 5 word-per-minute Morse code exam. You are introduced to each character and then you are given practice with that character. Each character is used in words or text before you move on to the next one. There is plenty of practice at 5 wpm to prepare you for your exam.

When you are ready to upgrade to the Amateur Extra license, *The ARRL Extra Class License Manual* will help you prepare. The book provides detailed explanations and examples of all types of calculations used on the exam. The complete Element 4 Extra class question pool is included.

Introduction

The General License

Earning a General class Amateur Radio license is a great way to increase your enjoyment of ham radio. The Element 3 written exam is straightforward, with no difficult math or electronics background required. You are sure to find the operating privileges available to a General class licensee to be worth the time spent learning about Amateur Radio. After passing the exam, you will be able to operate on every *band* that is assigned to the Amateur Radio Service. That means you will gain operating privileges on each of the HF bands, where world-wide communications is common. You'll be ready to enjoy voice and digital communications, as well as using the ham's own language — CW, or international Morse code. Experience the excitement of Amateur Radio!

Perhaps you are mainly interested in local communications using FM repeaters. Maybe you want to use your computer to explore the many digital modes of communication. If your eyes turn to the stars on a clear night, you might enjoy

Figure 1 — Rick Palm, K1CE, operating at ARRL's Hiram Percy Maxim Memorial Station, W1AW, enjoys a Morse code chat (QSO) with another ham.

tracking the amateur satellites and using them to relay your signals to other amateurs around the world!

Speaking of Morse code, there is a 5-words-per-minute Morse code exam for this license, along with the written exam. Many students are ready to pass the code exam after only a few weeks of study. Most people can pass that exam by the time they have learned all the required characters.

Once you make the commitment to study and learn what it takes to pass the exam, you *will* accomplish your goal. Many people pass the exam on their first try, so if you study the material and are prepared, chances are good that you will soon have your General license. It may take you more than one attempt to pass the General license exam, but that's okay. There is no limit to how many times you can take it. Many Volunteer Examiner Teams have several exam versions available, so you may even be able to try the exam again at the same exam session, if necessary. Time and available exam versions may limit the number of times you can try the exam at a single exam session. If you don't pass after a couple of tries you will certainly benefit from more study of the question pool before you try again.

An Overview of Amateur Radio

Earning an Amateur Radio license, at whatever level, is a special achievement. The 600,000 or so people in the US who call themselves Amateur Radio operators, or hams, are part of a global fraternity. Radio amateurs provide a voluntary, noncommercial, communication service. This is especially true during natural disasters or other emergencies. Hams have made many important contributions to the field of electronics and communications, and this tradition continues today. Amateur Radio experimentation is yet another reason many people become part of this self-disciplined group of trained operators, technicians and electronics experts — an asset to any country. Hams pursue their hobby purely for personal enrichment in technical and operating skills, without any type of payment except the personal satisfaction they feel from a job well done!

Radio signals do not know territorial boundaries, so hams have a unique ability to enhance international goodwill. Hams become ambassadors of their country every time they put their stations on the air.

Amateur Radio has been around since before World War I, and hams have always been at the forefront of technology. Today, hams relay signals through their own satellites, bounce signals off the moon, relay messages automatically through computerized radio networks and use any number of other "exotic" communications techniques. Amateurs talk from hand-held transceivers through mountaintop repeater stations that can relay their signals to other hams' cars or homes. Hams send their own pictures by television, talk with other hams around the world by voice or, keeping alive a distinctive traditional skill, tap out messages in Morse code. When emergencies arise, radio amateurs are on the spot to relay information to and from disaster-stricken areas that have lost normal lines of communication.

The US government, through the Federal Communications Commission (FCC), grants all US Amateur Radio licenses. This licensing procedure ensures operating skill and electronics know-how. Without this skill, radio operators, because of improperly adjusted equipment or neglected regulations, might unknowingly cause interference to other services using the radio spectrum.

Who Can Be a Ham?

The FCC doesn't care how old you are or whether you're a US citizen. If you pass the examination, the Commission will issue you an amateur license. Any person (except the agent of a foreign government) may take the exam and, if successful, receive an amateur license. It's important to understand that if a citizen of a foreign country receives an amateur license in this manner, he or she is a US Amateur Radio operator. (This should not be confused with alien reciprocal operation, which allows visitors from certain countries who hold valid amateur licenses in their homelands to operate their own stations in the US without having to take an FCC exam.)

License Structure

Anyone earning a new Amateur Radio license can earn one of three license classes — Technician, General and Amateur Extra. These vary in degree of knowledge required and frequency privileges granted. Higher class licenses have more comprehensive examinations. In return for passing a more difficult exam you earn more frequency privileges (frequency space and modes of operation). The vast majority of beginners start with the most basic license, the Technician, although it's possible to start with any class of license.

Technician licensees who learn the international Morse code and pass an exam to demonstrate their knowledge of code at 5 wpm gain some frequency privileges on four of the amateur high-frequency (HF) bands. This license was previously called the Technician Plus license, and many amateurs will refer to it by that name. **Table 1** lists the amateur license classes you can earn, along with a brief description of their exam requirements and operating privileges.

Although there are also other amateur license classes, the FCC is no longer issuing new licenses for these classes. The Novice license was long considered the beginner's license. Exams for this license were discontinued as of April 15, 2000. The FCC also stopped issuing new Advanced class licenses on that date. They will continue to renew previously issued licenses, however, so you will probably meet some Novice and Advanced class licensees on the air.

The written Technician exam, called Element 2, covers some basic radio fundamentals and knowledge of some of the rules and regulations in Part 97 of the FCC Rules.

Each step up the Amateur Radio license ladder requires the applicant to pass the lower exams. So to earn a General class or even an Amateur Extra class license, you must also pass the Technician written exam. This does not mean

Table 1
Amateur Operator Licenses†

Class	Code Test	Written Examination	Privileges
Technician		Basic theory and regulations. (Element 2)*	All amateur privileges above 50.0 MHz.
Technician with Morse code credit	5 wpm (Element 1)	Basic theory and regulations. (Element 2)*	All "Novice" HF privileges in addition to all Technician privileges.
General	5 wpm (Element 1)	Basic theory and regulations; General theory and regulations. (Elements 2 and 3)	All amateur privileges except those reserved for Advanced and Amateur Extra class.
Amateur Extra privileges.	5 wpm (Element 1)	All lower exam elements, plus Extra-class theory (Elements 2, 3 and 4)	All amateur

†A licensed radio amateur will be required to pass only those elements that are not included in the examination for the amateur license currently held.
*If you have a Technician-class license issued before March 21, 1987, you also have credit for Elements 1 and 3. You must be able to prove your Technician license was issued before March 21, 1987 to claim this credit.

you have to pass the Technician exam again if you already hold a Technician license! Your valid Amateur Radio license gives you credit for all the exam elements of that license when you go to upgrade. If you now hold a Technician license, you will only have to pass the Element 1 Morse code exam and the Element 3 General class written exam.

As a General, you can use a wide range of frequency bands — at least some portion of *all amateur bands*, in fact. You'll be able to talk with other hams around the world using voice, digital modes, slow-scan TV and Morse code on eight HF bands. (Another HF band — the 30-meter band — is reserved for Morse code and digital modes only.) See **Table 2**. You can provide public service through emergency communications and message handling.

You will experience the thrill of working (contacting) other Amateur Radio operators in just about any country in the world. There's nothing quite like making friends with other amateurs around the world.

Learning Morse Code

Even if you don't plan to use Morse code now, an international treaty requires it to earn those HF privileges. Learning Morse code is a matter of practice. Instructions on learning the code, how to handle a telegraph key, and so on, can

Table 2
Amateur Operating Privileges

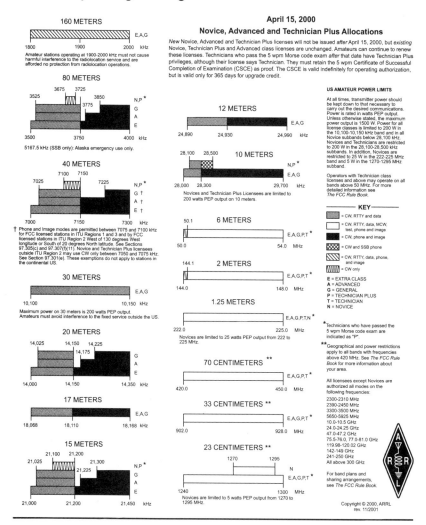

be found in *The ARRL General Class License Manual*. In addition, *Your Introduction to Morse Code*, ARRL's package to teach Morse code, is available with two cassette tapes or two audio CDs. *Your Introduction to Morse Code* was designed for beginners, and will help you learn the code and prepare to pass that 5-wpm code exam. You can purchase these products from your local Amateur Radio equipment dealer or directly from the ARRL, 225 Main St, Newington, CT 06111. To place an order, call, toll-free, **888-277-5289**. You can also send e-mail to: **pubsales@arrl.org** or check out our World Wide Web site:

Introduction 5

http://www.arrl.org/ Prospective new hams can call: **800-32-NEW HAM (800-326-3942)** for additional information.

Besides listening to code tapes or CDs, some on-the-air operating experience will be a great help in building your code speed. When you are in the middle of a contact via Amateur Radio, and have to copy the code the other station is sending to continue the conversation, your copying ability will improve quickly! Although you did not have to pass a Morse code test to earn your Technician license, there are no regulations prohibiting you from using code on the air. Many amateurs operate Morse code on the VHF and UHF bands.

ARRL's Maxim Memorial Station, W1AW, transmits code practice and information bulletins of interest to all amateurs. These code-practice sessions and Morse code bulletins provide an excellent opportunity for code practice. **Table 3** is a W1AW operating schedule. When we change from standard time to daylight saving time, the same local times are used.

Station Call Signs

Many years ago, by international agreement, the nations of the world decided to allocate certain call-sign prefixes to each country. This means that if you hear a radio station call sign beginning with W or K, for example, you know the station is licensed by the United States. A call sign beginning with the letter G

Figure 2—Once you begin operating on the amateur HF bands, you will soon begin to collect many colorful QSL cards. Many of them will likely be from countries around the world.

W1AW Schedule

PACIFIC	MTN	CENT	EAST	MON	TUE	WED	THU	FRI
6 AM	7 AM	8 AM	9 AM		FAST CODE	SLOW CODE	FAST CODE	SLOW CODE
7 AM-1 PM	8 AM-2 PM	9 AM-3 PM	10 AM-4 PM	VISITING OPERATOR TIME (12 PM - 1 PM CLOSED FOR LUNCH)			
1 PM	2 PM	3 PM	4 PM	FAST CODE	SLOW CODE	FAST CODE	SLOW CODE	FAST CODE
2 PM	3 PM	4 PM	5 PM	CODE BULLETIN				
3 PM	4 PM	5 PM	6 PM	TELEPRINTER BULLETIN				
4 PM	5 PM	6 PM	7 PM	SLOW CODE	FAST CODE	SLOW CODE	FAST CODE	SLOW CODE
5 PM	6 PM	7 PM	8 PM	CODE BULLETIN				
6 PM	7 PM	8 PM	9 PM	TELEPRINTER BULLETIN				
6:45 PM	7:45 PM	8:45 PM	9:45 PM	VOICE BULLETIN				
7 PM	8 PM	9 PM	10 PM	FAST CODE	SLOW CODE	FAST CODE	SLOW CODE	FAST CODE
8 PM	9 PM	10 PM	11 PM	CODE BULLETIN				

W1AW's schedule is at the same local time throughout the year. The schedule according to your local time will change if your local time does not have seasonal adjustments that are made at the same time as North American time changes between standard time and daylight time. From the first Sunday in April to the last Sunday in October, UTC = Eastern Time + 4 hours. For the rest of the year, UTC = Eastern Time + 5 hours.

♦ **Morse code transmissions:**
Frequencies are 1.818, 3.5815, 7.0475, 14.0475, 18.0975, 21.0675, 28.0675 and 147.555 MHz.
Slow Code = practice sent at 5, 7½, 10, 13 and 15 wpm.
Fast Code = practice sent at 35, 30, 25, 20, 15, 13 and 10 wpm.
Code practice text is from the pages of QST. The source is given at the beginning of each practice session and alternate speeds within each session. For example, "Text is from July 1992 QST, pages 9 and 81," indicates that the plain text is from the article on page 9 and mixed number/letter groups are from page 81.
Code bulletins are sent at 18 wpm.

W1AW qualifying runs are sent on the same frequencies as the Morse code transmissions. West Coast qualifying runs are transmitted on approximately 3.590 MHz by K6YR. See "Contest Corral" in this issue. At the beginning of each code practice session, the schedule for the next qualifying run is presented. Underline one minute of the highest speed you copied, certify that your copy was made without aid, and send it to ARRL for grading. Please include your name, call sign (if any) and complete mailing address. Send a 9×12-inch SASE for a certificate, or a business-size SASE for an endorsement.

♦ **Teleprinter transmissions:**
Frequencies are 3.625, 7.095, 14.095, 18.1025, 21.095, 28.095 and 147.555 MHz.
Bulletins are sent at 45.45-baud Baudot and 100-baud AMTOR, FEC Mode B. 110-baud ASCII will be sent only as time allows.
On Tuesdays and Fridays at 6:30 PM Eastern Time, Keplerian elements for many amateur satellites are sent on the regular teleprinter frequencies.

♦ **Voice transmissions:**
Frequencies are 1.855, 3.99, 7.29, 14.29, 18.16, 21.39, 28.59 and 147.555 MHz.

♦ **Miscellanea:**
On Fridays, UTC, a DX bulletin replaces the regular bulletins.
W1AW is open to visitors from 10 AM until noon and from 1 PM until 3:45 PM on Monday through Friday. FCC licensed amateurs may operate the station during that time. Be sure to bring your current FCC amateur license or a photocopy.
In a communication emergency, monitor W1AW for special bulletins as follows: voice on the hour, teleprinter at 15 minutes past the hour, and CW on the half hour.
Headquarters and W1AW are closed on New Year's Day, President's Day, Good Friday, Memorial Day, Independence Day, Labor Day, Thanksgiving and the following Friday, and Christmas Day.

is licensed by Great Britain, and a call sign beginning with VE is from Canada. *The ARRL DXCC List* is an operating aid no ham who is active on the HF bands should be without. That booklet, available from the ARRL, includes the common call-sign prefixes used by amateurs in virtually every location in the world. It also includes a check-off list to help you keep track of the countries you contact as you work toward collecting QSL cards from 100 or more countries to earn the prestigious DX Century Club award. (DX is ham lingo for distance, generally taken on the HF bands to mean any country outside the one from which you are operating.)

The International Telecommunication Union (ITU) radio regulations outline the basic principles used in forming amateur call signs. According to these regulations, an amateur call sign must be made up of one or two characters (the first one may be a numeral) as a prefix, followed by a numeral, and then a suffix of not more than three letters. The prefixes W, K, N and A are used in the United States. When the letter A is used in a US amateur call sign, it will always be with a two-letter prefix, AA to AL. The continental US is divided into 10 Amateur Radio call districts (sometimes called areas), numbered 0 through 9. **Figure 3** is a map showing the US call districts.

For information on the FCC's call-sign assignment system, and a table listing the blocks of call signs for each license class, see *The ARRL's FCC Rule Book*. You may keep the same call sign when you change license class, if you wish. You must indicate that you want to receive a new call sign when you fill out an FCC Form 605 to apply for the exam or change your address.

The FCC also has a vanity call sign system. Under this system the FCC will issue a call sign selected from a list of preferred available call signs. While

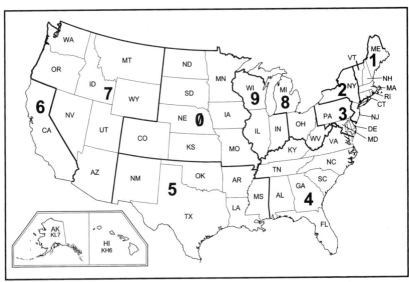

Figure 3—There are 10 US call areas. Hawaii is part of the sixth call area, and Alaska is part of the seventh.

there is no fee for an Amateur Radio license, there is a fee for the selection of a vanity call sign. The fee as of January, 2002 is $12 for a 10-year Amateur Radio license, paid upon application for a vanity call sign and at license renewal after that. (That fee may change as costs of administering the program change.) The latest details about the vanity call sign system are available from ARRL Regulatory Information, 225 Main Street, Newington, CT 06111-1494 and on ARRLWeb at **http://www.arrl.org/**

Earning a License

Forms and Procedures

To renew or modify a license, you can file a copy of FCC Form 605 or use one of the electronic filing options. Licenses are normally good for ten years. Your application for a license renewal must be submitted to the FCC no more than 90 days before the license expires. (We recommend you submit the application for renewal between 90 and 60 days before your license expires.) If the FCC receives your renewal application before the license expires, you may continue to operate until your new license arrives, even if it is past the expiration date.

If you forget to apply before your license expires, you may still be able to renew your license without taking another exam. There is a two-year grace period, during which you may apply for renewal of your expired license. Use an FCC Form 605 to apply for reinstatement (and your old call sign), or follow the electronic filing procedures. If you apply for reinstatement of your expired license under this two-year grace period, you may not operate your station until your new license is issued.

If you move or change addresses you should use an FCC Form 605 or electronic filing to notify the FCC of the change. If your license is lost or destroyed, however, just write a letter to the FCC explaining why you are requesting a new copy of your license.

You can ask one of the Volunteer Examiner Coordinators' offices to file your renewal application electronically if you don't want to mail the form to the FCC. You must still mail the form to the VEC, however. The ARRL/VEC Office will electronically file application forms for any ARRL member free of charge.

Electronic Filing

You can also file your license renewal or address modification using the Universal Licensing System (ULS) on the World Wide Web. To use ULS, you must have an FCC Registration Number, or FRN. Obtain your FRN by registering with the Commission Registration System, known as CORES.

Described as an agency-wide registration system for anyone filing applications with or making payments to the FCC, CORES will assign a unique 10-digit FCC Registration Number, or FRN to all registrants. All Commission systems that handle financial, authorization of service, and enforcement activities will use the FRN. The FCC says use of the FRN will allow it to more rapidly verify fee payment. Amateurs mailing payments to the FCC — for example as part of a vanity call sign application — would include their FRN on FCC Form 159.

The on-line filing system and further information about CORES is available by visiting the FCC Web site, **http://www.fcc.gov** and clicking on the Commission Registration System link. Follow the directions on the Web site. It is also possible to register on CORES using a paper Form 160.

When you register with CORES you must supply a Taxpayer Identification Number, or TIN. For individuals, this is usually a Social Security Number. Club stations must obtain an Assigned Taxpayer Identification Number (ATIN) before registering on CORES

Anyone can register on CORES and obtain an FRN. CORES/FRN is "entity registration." You don't need a license to be registered.

Once you have registered on CORES and obtained your FRN, you can proceed to renew or modify your license using the Universal Licensing System (ULS), also on the World Wide Web. Go to **http://www.fcc.gov/wtb/uls** and click on the "Online Filing" button. Follow the directions provided on the Web page to connect to the FCC's ULS database.

Paper Filing

The FCC has a set of detailed instructions for the Form 605, which are included with the form. To obtain a new Form 605, call the FCC Forms Distribution Center at 800-418-3676. You can also write to: Federal Communications Commission, Forms Distribution Center, 2803 52nd Avenue, Hyattsville, MD 20781 (specify "Form 605" on the envelope). The Form 605 also is available from the FCC's fax on demand service. Call 202-418-0177 and ask for form number 000605. Form 605 also is available via the Internet. The World Wide Web location is: **http://www.fcc.gov/formpage.html** or you can receive the form via ftp to: **ftp.fcc.gov/pub/Forms/Form605**.

The ARRL/VEC has created a package that includes the portions of Form 605 that are needed for amateur applications, as well as a condensed set of instructions for completing the form. Write to: ARRL/VEC, Form 605, 225 Main Street, Newington, CT 06111-1494. (Please include a large business-sized stamped self-addressed envelope with your request.) **Figure 4** is a sample of those portions of an FCC Form 605 that you would complete to submit a change of address to the FCC.

Most of the form is simple to fill out. You will need to know that the Radio Service Code for box 1 is HA for Amateur Radio. (Just remember HAm radio.) You will have to include a "Taxpayer Identification Number" on the Form. This is normally your Social Security Number. If you don't want to write your Social Security Number on this form, then you can register with CORES as described above. Then you will receive your FRN from the FCC, and you can use that number instead of your Social Security Number on the Form. Of course, you will have to supply your Social Security Number to register with the CORES.

The telephone number, fax number and e-mail address information is optional. The FCC will use that information to contact you in case there is a problem with your application.

Page two of the Form includes six General Certification Statements. Statement five may seem confusing. Basically, this statement means that you do not plan to install an antenna over 200 feet high, and that your permanent station location will not be in a designated wilderness area, wildlife preserve or nationally recognized scenic and recreational area.

The sixth statement indicates that you are familiar with the FCC RF Safety Rules, and that you will obey them. Chapter 10 (Subelement G0) includes exam questions and explanations about the RF Safety Rules.

Volunteer Examiner Program

Before you can take an FCC exam, you'll have to fill out a copy of the National Conference of Volunteer Examiner Coordinators' (NCVEC) Quick Form 605. This form is used as an application for a new license or an upgraded license. The NCVEC Quick Form 605 is only used at license exam sessions. This form includes some information that the Volunteer Examiner Coordinator's office will need to process your application with the FCC. See **Figure 5**. You should not use an NCVEC Quick Form 605 to apply for a license renewal or modification with the FCC. Never mail these forms to the FCC, because that will result in a rejection of the application. Likewise, an FCC Form 605 can't be used for an exam application.

All US amateur exams are administered by Volunteer Examiners who are certified by a Volunteer-Examiner Coordinator (VEC). Program. *The ARRL's FCC Rule Book* contains more details about the Volunteer-Examiner program.

To qualify for a General license you must pass Morse code Element 1 and written Elements 2 and 3. If you already hold a valid Novice license, or a Technician Plus license (which includes Morse code credit), then you have credit for passing Element 1. If you have a Technician license you will receive credit for Element 2. In that case, you will only have to pass the Element 3 written exam to complete your upgrade to General. See **Table 4** for a summary of the Exam Elements you will need to qualify for a General class license. It is an unfortunate quirk of the FCC Rules that if you currently have a Technician license and then pass the Morse code exam, your Certificate of Successful Completion of Examination (CSCE) is valid for 365 days toward your General class upgrade. After that, you will have to take and pass the Morse code exam again to complete your upgrade. The CSCE for the code credit is valid indefinitely to use the Novice/Technician with Morse code privileges on the HF bands, but only valid for 365 days to upgrade to General.

The Element 3 exam consists of 35 questions taken from a pool of more than 350. The question pools for all amateur exams are maintained by a Question PoolCommittee selected by the Volunteer Examiner Coordinators. The FCC allows Volunteer Examiners to select the questions for an amateur exam, but they must use the questions exactly as they are released by the VEC that coordinates the test session. If you attend a test session coordinated by the ARRL/VEC, your test will be designed by the ARRL/VEC or by a computer program designed by the VEC. The questions and answers will be exactly as they are printed in this book. Be careful, though. The ARRL/VEC and some other VECs scramble the positions of the answers within the questions.

FCC 605 Main Form	Quick-Form Application for Authorization in the Ship, Aircraft, Amateur, Restricted and Commercial Operator, and General Mobile Radio Services	Approved by OMB 3060 - 0850 See instructions for public burden estimate

1) Radio Service Code: **HA**

Application Purpose (Select only one) (M **D**)

2)
- NE – New
- MD – Modification
- AM – Amendment
- RO – Renewal Only
- RM – Renewal / Modification
- CA – Cancellation of License
- WD – Withdrawal of Application
- DU – Duplicate License
- AU – Administrative Update

3) If this request is for Developmental License or STA (Special Temporary Authorization) enter the appropriate code and attach the required exhibit as described in the instructions. Otherwise enter N (Not Applicable). — (**N**) D S N/A

4) If this request is for an Amendment or Withdrawal of Application, enter the file number of the pending application currently on file with the FCC. — File Number

5) If this request is for a Modification, Renewal Only, Renewal / Modification, Cancellation of License, Duplicate License, or Administrative Update, enter the call sign (serial number for Commercial Operator) of the existing FCC license. If this is a request for consolidation of DO & DM Operator Licenses, enter serial number of DO. — Call Sign/Serial # **W R 1 B**

6) If this request is for a New, Amendment, Renewal Only, or Renewal Modification, enter the requested expiration date of the authorization (this item is optional). — MM DD

7) Does this filing request a Waiver of the Commission's rules? If 'Y', attach the required showing as described in the instructions. — ( ) Yes **No**

8) Are attachments (other than associated schedules) being filed with this application? — ( ) Yes **No**

Applicant Information

9) FCC Registration Number (FRN): **0003-3573-99**

10) Applicant /Licensee is a(n): (**I**) Individual Unincorporated Association Trust Government Entity Joint Venture
Corporation Limited Liability Corporation Partnership Consortium

11) First Name (If individual): **Larry**   MI: **D**   Last Name: **Wolfgang**   Suffix:

11a) Date of Birth (required for Amateur Radio or Commercial Operators (including Restricted Radiotelephone)): **07** (mm)/ **09** (dd)/ **52** (yy)

12) Entity Name (if other than individual):

13) Attention To:

14) P.O. Box: And/Or   15) Street Address: **225 Main Street**

16) City: **Newington**   17) State: **CT**   18) Zip: **06111**   19) Country: **USA**

20) Telephone Number: **860-594-0200**   21) FAX:

22) E-Mail Address: **wr1b@arrl.net**

FCC 605 – Main Form
November 2001 - Page 1

Figure 4 — This sample FCC Form 605 shows the sections you should complete to notify the FCC by mail of a change in your address.

Fee Status

23)	Is the applicant exempt from FCC application Fees?	(N) Yes No
24)	Is the applicant exempt from FCC regulatory Fees?	(N) Yes No

General Certification Statements

1) The Applicant waives any claim to the use of any particular frequency or of the electromagnetic spectrum as against the regulatory power of the United States because of the previous use of the same, whether by license or otherwise, and requests an authorization in accordance with this application.
2) The applicant certifies that all statements made in this application and in the exhibits, attachments, or documents incorporated by reference are material, are part of this application, and are true, complete, correct, and made in good faith.
3) Neither the Applicant nor any member thereof is a foreign government or a representative thereof.
4) The applicant certifies that neither the applicant nor any other party to the application is subject to a denial of Federal benefits pursuant to Section 5301 of the Anti-Drug Abuse Act of 1988, 21 U.S.C. § 862, because of a conviction for possession or distribution of a controlled substance. **This certification does not apply to applications filed in services exempted under Section 1.2002(c) of the rules, 47 CFR § 1.2002(c).** See Section 1.2002(b) of the rules, 47 CFR § 1.2002(b), for the definition of "party to the application" as used in this certification.
5) Amateur or GMRS Applicant certifies that the construction of the station would NOT be an action which is likely to have a significant environmental effect (see the Commission's Rules 47 CFR Sections 1.1301-1.1319 and Section 97.13(a) Rules (available at web site http://www.fcc.gov/wtb/rules.html).
6) Amateur Applicant certifies that they have READ and WILL COMPLY WITH Section 97.13(c) of the Commission's Rules (available at web site http://www.fcc.gov/wtb/rules.html) regarding RADIOFREQUENCY (RF) RADIATION SAFETY and the amateur service section of OST/OET Bulletin Number 65 (available at web site http://www.fcc.gov/oet/info/documents/bulletins/).

Certification Statements For GMRS Applicants

1) Applicant certifies that he or she is claiming eligibility under Rule Section 95.5 of the Commission's Rules.
2) Applicant certifies that he or she is at least 18 years of age.
3) Applicant certifies that he or she will comply with the requirement that use of frequencies 462.650, 467.650, 462.700 and 467.700 MHz is not permitted near the Canadian border North of Line A and East of Line C. These frequencies are used throughout Canada and harmful interference is anticipated.
4) Non-Individual applicants certify that they have NOT changed frequency or channel pairs, type of emission, antenna height, location of fixed transmitters, number of mobile units, area of mobile operation, or increase in power.

Signature

25) Typed or Printed Name of Party Authorized to Sign

First Name: Larry	MI: D	Last Name: Wolfgang	Suffix:

26) Title:

Signature: Larry D. Wolfgang	27) Date: 1/2/2002

Failure to Sign This Application May Result in Dismissal Of The Application And Forfeiture Of Any Fees Paid

WILLFUL FALSE STATEMENTS MADON THIS FOR OR ANY ATTACHMENTS ARE PUNISHBABLE BY FINE AND/OR IMPRISONMENT (U.S. Code, Title 18, Section 1001) AND / OR REVOCATION OF ANY STATION LICENSE OR CONSTRUCTION PERMIT (U.S. Code, Title 47, Section 312(a)(1)), AND / OR FORFEITURE (U.S. Code, Title 47, Section 503).

Introduction 13

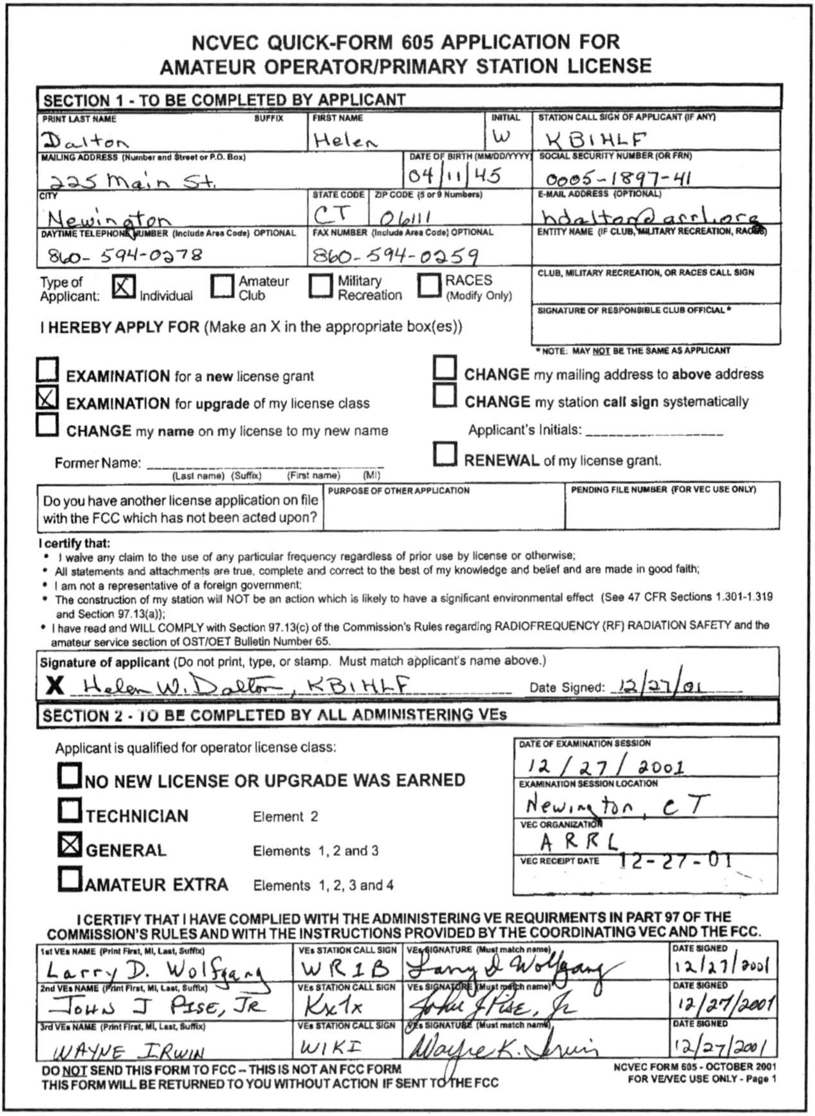

NCVEC QUICK-FORM 605 APPLICATION FOR
AMATEUR OPERATOR/PRIMARY STATION LICENSE

SECTION 1 - TO BE COMPLETED BY APPLICANT

PRINT LAST NAME / SUFFIX	FIRST NAME	INITIAL	STATION CALL SIGN OF APPLICANT (IF ANY)
Dalton	Helen	W	KB1HLF

MAILING ADDRESS (Number and Street or P.O. Box)	DATE OF BIRTH (MM/DD/YYYY)	SOCIAL SECURITY NUMBER (OR FRN)
225 Main St.	04/11/45	0005-1897-41

CITY	STATE CODE	ZIP CODE (5 or 9 Numbers)	E-MAIL ADDRESS (OPTIONAL)
Newington	CT	06111	hdalton@arrl.org

DAYTIME TELEPHONE NUMBER (Include Area Code) OPTIONAL	FAX NUMBER (Include Area Code) OPTIONAL	ENTITY NAME (IF CLUB, MILITARY RECREATION, RACES)
860-594-0278	860-594-0259	

Type of Applicant: [X] Individual [ ] Amateur Club [ ] Military Recreation [ ] RACES (Modify Only)

CLUB, MILITARY RECREATION, OR RACES CALL SIGN

SIGNATURE OF RESPONSIBLE CLUB OFFICIAL*

*NOTE: MAY NOT BE THE SAME AS APPLICANT

I HEREBY APPLY FOR (Make an X in the appropriate box(es))

[ ] EXAMINATION for a new license grant
[X] EXAMINATION for upgrade of my license class
[ ] CHANGE my name on my license to my new name
Former Name: ____ (Last name) (Suffix) (First name) (MI)
[ ] CHANGE my mailing address to above address
[ ] CHANGE my station call sign systematically
Applicant's Initials: ____
[ ] RENEWAL of my license grant.

Do you have another license application on file with the FCC which has not been acted upon?	PURPOSE OF OTHER APPLICATION	PENDING FILE NUMBER (FOR VEC USE ONLY)

I certify that:
- I waive any claim to the use of any particular frequency regardless of prior use by license or otherwise;
- All statements and attachments are true, complete and correct to the best of my knowledge and belief and are made in good faith;
- I am not a representative of a foreign government;
- The construction of my station will NOT be an action which is likely to have a significant environmental effect (See 47 CFR Sections 1.301-1.319 and Section 97.13(a));
- I have read and WILL COMPLY with Section 97.13(c) of the Commission's Rules regarding RADIOFREQUENCY (RF) RADIATION SAFETY and the amateur service section of OST/OET Bulletin Number 65.

Signature of applicant (Do not print, type, or stamp. Must match applicant's name above.)
X Helen W. Dalton, KB1HLF Date Signed: 12/27/01

SECTION 2 - TO BE COMPLETED BY ALL ADMINISTERING VEs

Applicant is qualified for operator license class:
[ ] NO NEW LICENSE OR UPGRADE WAS EARNED
[ ] TECHNICIAN Element 2
[X] GENERAL Elements 1, 2 and 3
[ ] AMATEUR EXTRA Elements 1, 2, 3 and 4

DATE OF EXAMINATION SESSION: 12/27/2001
EXAMINATION SESSION LOCATION: Newington, CT
VEC ORGANIZATION: ARRL
VEC RECEIPT DATE: 12-27-01

I CERTIFY THAT I HAVE COMPLIED WITH THE ADMINISTERING VE REQUIRMENTS IN PART 97 OF THE COMMISSION'S RULES AND WITH THE INSTRUCTIONS PROVIDED BY THE COORDINATING VEC AND THE FCC.

VEs NAME (Print First, MI, Last, Suffix)	VEs STATION CALL SIGN	VEs SIGNATURE (Must match name)	DATE SIGNED
1st: Larry D. Wolfgang	WR1B	Larry D. Wolfgang	12/27/2001
2nd: John J Pise, Jr	Kx1x	John J Pise, Jr	12/27/2001
3rd: Wayne Irwin	W1KI	Wayne K. Irwin	12/27/2001

DO NOT SEND THIS FORM TO FCC – THIS IS NOT AN FCC FORM
THIS FORM WILL BE RETURNED TO YOU WITHOUT ACTION IF SENT TO THE FCC

NCVEC FORM 605 - OCTOBER 2001
FOR VE/VEC USE ONLY - Page 1

Figure 5 — A completed NCVEC Quick Form 605 as it would be filled out for a new General class license.

Finding an Exam Opportunity

To determine where and when exams will be given in your area, contact the ARRL/VEC office, or watch for announcements in the Hamfest Calendar and Coming Conventions columns in *QST*. Many local clubs sponsor exams, so they are another good source of information on exam opportunities. Upcoming exams

Table 4
Exam Elements Needed to Qualify for a General Class License

Current License	Exam Requirements	Study Materials
None	Morse code (Element 1) Technician (Element 2)	Your Introduction to Morse Code Now You're Talking!, 4th Ed or ARRL Technician Class Video Course or ARRL's Tech Q & A
	General (Element 3)	The ARRL General Class License Manual or ARRL General Class Video Course or ARRL's General Q & A
Novice	Technician (Element 2)	Now You're Talking!, 4th Ed or ARRL Technician Class Video Course or ARRL's Tech Q & A
	General (Element 3)	The ARRL General Class License Manual or ARRL General Class Video Course or ARRL's General Q & A
Technician issued on or after Feb 14, 1991	Morse code (Element 1) General (Element 3)	Your Introduction to Morse Code The ARRL General Class License Manual or ARRL General Class Video Course or ARRL's General Q & A
Technician issued before Feb 14, 1991	General (Element 3)	The ARRL General Class License Manual ARRL General Class Video Course-ARRL's General Q & A
Technician Plus or Technician with Morse code credit	General (Element 3)	The ARRL General Class License Manual or ARRL General Class Video Course or ARRL's General Q & A
Technician issued before Mar 21, 1987	None*	

*Individuals who qualified for the Technician license before March 21, 1987 will be able to upgrade to General class by providing documentary proof to a Volunteer Examiner Coordinator, paying an application fee and completing NCVEC Quick Form 605.

---

are listed on *ARRLWeb* at: **http://www.arrl.org/arrlvec/examsearch.phtml**. Registration deadlines, and the time and location of the exams, are mentioned prominently in publicity releases about upcoming sessions.

Taking the Exam

By the time examination day rolls around, you should have already prepared yourself. This means getting your schedule, supplies and mental attitude ready.

Plan your schedule so you'll get to the examination site with plenty of time to spare. There's no harm in being early. In fact, you might have time to discuss hamming with another applicant, which is a great way to calm pretest nerves. Try not to discuss the material that will be on the examination, as this may make you even more nervous. By this time, it's too late to study anyway!

What supplies will you need? First, be sure you bring your current original Amateur Radio license, if you have one. Bring a photocopy of your license, too, as well as the original and a photocopy of any Certificates of Successful Completion of Examination (CSCE) that you plan to use for exam credit. Bring along several sharpened number 2 pencils and two pens (blue or black ink). Be sure to have a good eraser. A pocket calculator may also come in handy. You may use a programmable calculator if that is the kind you have, but take it into your exam "empty" (cleared of all programs and constants in memory). Don't program equations ahead of time, because you may be asked to demonstrate that there is nothing in the calculator memory. The examining team has the right to refuse a candidate the use of any calculator that they feel may contain information for the test or could otherwise be used to cheat on the exam.

The Volunteer Examiner Team is required to check two forms of identification before you enter the test room. This includes your *original* Amateur Radio license, if you have one — not a photocopy. (You will need a photocopy of your license to file with your application, but only the original is valid for ID.) A photo ID of some type is best for the second form of ID, but is not required by the FCC. Other acceptable forms of identification include a driver's license, a piece of mail addressed to you or a birth certificate.

The following description of the testing procedure applies to exams coordinated by the ARRL/VEC, although many other VECs use a similar procedure.

Code Test

The code test is usually given before the written exams. If you don't plan to take the code exam, just sit quietly while the other candidates give it a try.

Before you take the code test, you'll be handed a piece of paper to copy the code as it is sent. The test will begin with about a minute of practice copy. Then comes the actual test: at least five minutes of Morse code. You are responsible for knowing the 26 letters of the alphabet, the numerals 0 through 9, the period, comma, question mark, and the procedural signals $\overline{AR}$(+), $\overline{SK}$, $\overline{BT}$ (= or double dash) and $\overline{DN}$ (/ or fraction bar, sometimes called the "slant bar").

You may copy the entire text word for word, or just take notes on the content. At the end of the transmission, the examiner will hand you 10 questions about the text. Fill in the blanks with your answers. (You must spell each answer exactly as it was sent.) If you get at least 7 correct, you pass! Alternatively, the exam team has the option to look at your copy sheet if you fail the 10-question exam. If you have one minute of solid copy (25 characters), the examiners can certify that you passed the test on that basis. The format of the test transmission is generally similar to one side of a normal on-the-air amateur conversation (QSO).

A sending test may not be required. The Commission has decided that if applicants can demonstrate receiving ability, they most likely can also send at that speed. But be prepared for a sending test, just in case! Subpart 97.503(a) of the FCC Rules says, "A telegraphy examination must be sufficient to prove that the examinee has the ability to send correctly by hand and to receive correctly by ear texts in the international Morse code at not less than the prescribed speed..."

Written Tests

After the code tests are administered, you'll take the written examination. The examiner will give each applicant a test booklet, an answer sheet and scratch paper. After that, you're on your own. The first thing to do is read the instructions. Be sure to sign your name every place it's called for. Do all of this at the beginning to get it out of the way.

Next, check the examination to see that all pages and questions are there. If not, report this to the examiner immediately. When filling in your answer sheet make sure your answers are marked next to the numbers that correspond to each question.

Go through the entire exam, and answer the easy questions first. Next, go back to the beginning and try the harder questions. Leave the really tough questions for last. Guessing can only help, as there is no additional penalty for answering incorrectly.

If you have to guess, do it intelligently: At first glance, you may find that you can eliminate one or more "distracters." Of the remaining responses, more than one may seem correct; only one is the best answer, however. To the applicant who is fully prepared, incorrect distracters to each question are obvious. Nothing beats preparation!

After you've finished, check the examination thoroughly. You may have read a question wrong or goofed in your arithmetic. Don't be overconfident. There's no rush, so take your time. Think, and check your answer sheet. When you feel you've done your best and can do no more, return the test booklet, answer sheet and scratch pad to the examiner.

The Volunteer-Examiner Team will grade the exam while you wait. The passing mark is 74%. (That means 26 out of 35 questions correct — or no more than 9 incorrect answers on the Element 3 exam.) You will receive a Certificate of Successful Completion of Examination (CSCE) showing all exam elements that you pass at that exam session. If you are already licensed, and you pass the exam elements required to earn a higher license class, the CSCE authorizes you to operate with your new privileges immediately. When you use these new privileges, you must sign your call sign followed by the slant mark ("/"; on voice, say "stroke" or "slant") and the letters "AG," if you are upgrading from a Novice or Technician to a General class license. You only have to follow this special identification procedure until your new license is granted by the FCC, however.

If you pass only some of the exam elements required for a license, you will still receive a CSCE. That certificate shows what exam elements you passed, and is valid for 365 days. Use it as proof that you passed those exam elements so you won't have to take them over again next time you try for the license.

And Now, Let's Begin

The complete General question pool (Element 3) is printed in this book. Each chapter lists all the questions for a particular subelement (such as Electrical Principles — G5). A brief explanation about the correct answer is given after each question.

Table 5 shows the study guide or syllabus for the Element 3 exam as released by the Volunteer-Examiner Coordinators' Question Pool Committee in December 1999. The syllabus lists the topics to be covered by the General class exam, and so forms the basic outline for the remainder of this book. Use the syllabus to guide your study, and to ensure that you have studied the material for all of the topics listed.

The question numbers used in the question pool refer to this syllabus. Each question number begins with a syllabus-point number (for example, G1E or G0C). The question numbers end with a two-digit number. For example, question G3B09 is the ninth question about the G3B syllabus point.

The Question Pool Committee designed the syllabus and question pool so there are the same number of points in each subelement as there are exam questions from that subelement. For example, three exam questions on the General exam must be from the "Radio-Wave Propagation" subelement, so there are three groups for that point. These are numbered G3A, G3B and G3C. While not a requirement of the FCC Rules, the Question Pool Committee recommends that one question be taken from each group to make the best possible license exams.

Good luck with your studies!

Table 5
General (Element 3) Syllabus

SUBELEMENT G1 — COMMISSION'S RULES
[6 Exam Questions — 6 Groups]

G1A General control operator frequency privileges
G1B Antenna structure limitations; good engineering and good amateur practice; beacon operation; restricted operation; retransmitting radio signals
G1C Transmitter power standards
G1D Examination element preparation; examination administration; temporary station identification
G1E Local control; repeater and harmful interference definitions; third party communications
G1F Certification of external RF-power-amplifiers; standards for certification of external RF-power amplifiers; HF data emission standards

SUBELEMENT G2 — OPERATING PROCEDURES
[6 Exam Questions — 6 Groups]

G2A Phone operating procedures
G2B Operating courtesy
G2C Emergencies, including drills and emergency communications
G2D Amateur auxiliary to the FCC's Compliance and Information Bureau; antenna orientation to minimize interference; HF operations, including logging practices
G2E Third-party communications; ITU Regions; VOX operation
G2F CW operating procedures, including procedural signals, Q signals and common abbreviations; full break-in; RTTY operating procedures, including procedural signals and common abbreviations and operating procedures for other digital modes, such as HF packet, AMTOR, PacTOR, G-TOR, Clover and PSK31

SUBELEMENT G3 — RADIO WAVE PROPAGATION
[3 Exam Questions — 3 Groups]

G3A Ionospheric disturbances; sunspots and solar radiation
G3B Maximum usable frequency; propagation "hops"
G3C Height of ionospheric regions; critical angle and frequency; HF scatter

SUBELEMENT G4 — AMATEUR RADIO PRACTICES
[5 Exam Questions — 5 Groups]

G4A Two-tone test; electronic TR switch; amplifier neutralization
G4B Test equipment: oscilloscope; signal tracer; antenna noise bridge; monitoring oscilloscope; field-strength meters
G4C Audio rectification in consumer electronics; RF ground
G4D Speech processors; PEP calculations; wire sizes and fuses
G4E Common connectors used in amateur stations: types; when to use; fastening methods; precautions when using; HF mobile radio installations; emergency power systems; generators; battery storage devices and charging sources including solar; wind generation

SUBELEMENT G5 — ELECTRICAL PRINCIPLES
[2 Exam Questions — 2 Groups]

G5A Impedance; including matching; resistance; including ohm; reactance; inductance; capacitance and metric divisions of these values

G5B Decibel; Ohm's Law; current and voltage dividers; electrical power calculations and series and parallel components; transformers (either voltage or impedance); sine wave root-mean-square (RMS) value

SUBELEMENT G6 — CIRCUIT COMPONENTS
[1 Exam question — 1 Group]

G6A Resistors; capacitors; inductors; rectifiers and transistors; etc.

SUBELEMENT G7 — PRACTICAL CIRCUITS
[1 Exam question — 1 Group]

G7A Power supplies and filters; single-sideband transmitters and receivers

SUBELEMENT G8 — SIGNALS AND EMISSIONS
[2 Exam Questions — 2 Groups]

G8A Signal information; AM; FM; single and double sideband and carrier; bandwidth; modulation envelope; deviation; overmodulation

G8B Frequency mixing; multiplication; bandwidths; HF data communications

SUBELEMENT G9 — ANTENNAS AND FEED-LINES
[4 Exam Questions — 4 Groups]

G9A Yagi antennas - physical dimensions; impedance matching; radiation patterns; directivity and major lobes

G9B Loop antennas - physical dimensions; impedance matching; radiation patterns; directivity and major lobes

G9C Random wire antennas - physical dimensions; impedance matching; radiation patterns; directivity and major lobes; feed point impedance of $1/2$-wavelength dipole and $1/4$-wavelength vertical antennas

G9D Popular antenna feed-lines - characteristic impedance and impedance matching; SWR calculations

SUBELEMENT G0 — RF SAFETY
[5 Exam Questions — 5 Groups]

G0A RF Safety Principles
G0B RF Safety Rules and Guidelines
G0C Routine Station Evaluation and Measurements (FCC Part 97 refers to RF Radiation Evaluation)
G0D Practical RF-safety applications
G0E RF-safety solutions

Subelement G1

Commission's Rules

Your General class (Element 3) written exam will consist of 35 questions, taken from the General class question pool. This question pool is prepared by the Volunteer Examiner Coordinators' Question Pool Committee. A certain number of questions are taken from each of the 10 subelements. There will be 6 questions from the Commission's Rules subelement printed in this chapter. These questions are divided into 6 groups, labeled G1A through G1F.

After most of the explanations in this chapter you will see a reference to Part 97 of the FCC Rules set inside square brackets, like [97.301d]. This tells you where to look for the exact wording of the Rules as they relate to that question. For a complete copy of Part 97, along with simple explanations of the Rules governing Amateur Radio, see *The FCC Rule Book* published by ARRL.

G1A General control operator frequency privileges

G1A01 What are the frequency limits for General class operators in the 160-meter band (ITU Region 2)?
- A. 1800 - 1900 kHz
- B. 1900 - 2000 kHz
- C. 1800 - 2000 kHz
- D. 1825 - 2000 kHz

C General class operators are allowed CW, RTTY, data, phone, and image privileges in the entire 160-meter band, from 1800 to 2000 kHz. ITU stands for International Telecommunication Union. This is the international organization that allocates frequencies among various types of users (commercial broadcast, government, land mobile, and so on.). ITU Region 2 includes North and South America. [97.301(d)]

160 METERS
CW, RTTY, data, phone, and image

1800 — 1900 — 2000 kHz   G

Amateur stations operating at 1900-2000 kHz must not cause harmful interference to the radiolocation service and are afforded no protection from radiolocation operations.

G1A02 What are the frequency limits for General class operators in the 75/80-meter band (ITU Region 2)?
A. 3525 - 3750 kHz and 3850 - 4000 kHz
B. 3525 - 3775 kHz and 3875 - 4000 kHz
C. 3525 - 3750 kHz and 3875 - 4000 kHz
D. 3525 - 3775 kHz and 3850 - 4000 kHz

A   General class operators have privileges on two parts of the 75/80-meter band. Between 3525 and 3750 kHz General class operators have CW, RTTY, and data privileges. Between 3850 and 4000 kHz General class operators have CW, phone, and image privileges. ITU stands for International Telecommunication Union. ITU Region 2 includes North and South America. [97.301(d)]

80 METERS

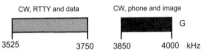

G1A03 What are the frequency limits for General class operators in the 40-meter band (ITU Region 2)?
A. 7025 - 7175 kHz and 7200 - 7300 kHz
B. 7025 - 7175 kHz and 7225 - 7300 kHz
C. 7025 - 7150 kHz and 7200 - 7300 kHz
D. 7025 - 7150 kHz and 7225 - 7300 kHz

D   General class operators have privileges on two parts of the 40-meter band. General class operators have CW, RTTY, and data privileges between 7025 and 7150 kHz and they have CW, phone, and image privileges between 7225 and 7300 kHz. ITU stands for International Telecommunication Union. [97.301(d)]

40 METERS

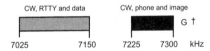

† Phone and Image modes are permitted between 7075 and 7100 kHz for FCC licensed stations in ITU Regions 1 and 3 and by FCC licensed stations in ITU Region 2 West of 130 degrees West longitude or South of 20 degrees North latitude. See Sections 97.305(c) and 97.307(f)(11). Novice and Technician Plus licensees outside ITU Region 2 may use CW only between 7050 and 7075 kHz. See Section 97.301(e). These exemptions do not apply to stations in the continental US.

G1A04 What are the frequency limits for General class operators in the 30-meter band?
- A. 10100 - 10150 kHz
- B. 10100 - 10175 kHz
- C. 10125 - 10150 kHz
- D. 10125 - 10175 kHz

A General class operators have CW, RTTY, and data privileges available to them throughout the 30-meter amateur band: 10100 to 10150 kHz. [97.301(d)]

30 METERS

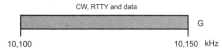

Maximum power on 30 meters is 200 watts PEP output.
Amateurs must avoid interference to the fixed service outside the US.

G1A05 What are the frequency limits for General class operators in the 20-meter band?
- A. 14025 - 14100 kHz and 14175 - 14350 kHz
- B. 14025 - 14150 kHz and 14225 - 14350 kHz
- C. 14025 - 14125 kHz and 14200 - 14350 kHz
- D. 14025 - 14175 kHz and 14250 - 14350 kHz

B General class operators have CW, RTTY, and data privileges from 14025 to 14150 kHz and they have CW, phone, and image privileges from 14225 to 14350 kHz. [97.301(d)]

20 METERS

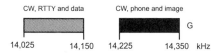

Commission's Rules 23

G1A06 What are the frequency limits for General class operators in the 15-meter band?
- A. 21025 - 21200 kHz and 21275 - 21450 kHz
- B. 21025 - 21150 kHz and 21300 - 21450 kHz
- C. 21025 - 21150 kHz and 21275 - 21450 kHz
- D. 21025 - 21200 kHz and 21300 - 21450 kHz

D   General class operators in the 15-meter band have CW, RTTY, and data privileges between 21025 and 21200 kHz and they have CW, phone, and image privileges between 21300 and 21450 kHz. [97.301(d)]

15 METERS

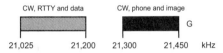

G1A07 What are the frequency limits for General class operators in the 12-meter band?
- A. 24890 - 24990 kHz
- B. 24890 - 24975 kHz
- C. 24900 - 24990 kHz
- D. 24900 - 24975 kHz

A   General class operators have access to the entire 12-meter band, from 24890 to 24990 kHz. Different privileges are allowed on two sections of the band. General class operators have CW, RTTY, and data privileges from 24890 to 24930 kHz and they have CW, phone, and image privileges from 24930 to 24990 kHz. [97.301(d)]

12 METERS

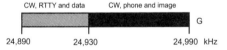

G1A08 What are the frequency limits for General class operators in the 10-meter band?
   A. 28000 - 29700 kHz
   B. 28025 - 29700 kHz
   C. 28100 - 29600 kHz
   D. 28125 - 29600 kHz

A   General class operators have access to the entire 10-meter band, from 28000 to 29700 kHz. Different privileges are allowed on two sections of the band. From 28000 to 28300 General class operators have CW, RTTY, and data privileges and from 28300 to 29700 kHz they have CW, phone, and image privileges. [97.301(d)]

G1A09 What are the frequency limits for General class operators in the 17-meter band?
   A. 18068 - 18300 kHz
   B. 18025 - 18200 kHz
   C. 18100 - 18200 kHz
   D. 18068 - 18168 kHz

D   General class operators have access to the entire 17-meter band, from 180680 to 18168 kHz. Different privileges are allowed on two sections of the band. Between 18068 and 18110 General class operators have CW, RTTY, and data privileges and from 18110 to 18168 kHz they have CW, phone, and image privileges. [97.301(d)]

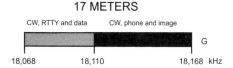

G1A10 What class of amateur license authorizes you to operate on the frequencies 14025 - 14150 kHz and 14225 - 14350 kHz?
A. Amateur Extra class only
B. Amateur Extra, Advanced or General class
C. Amateur Extra, Advanced, General or Technician class
D. Amateur Extra and Advanced class only

B General class operators may operate CW, RTTY and data between 14025 and 14150 kHz. They may also operate CW, phone and image between 14225 and 14350 kHz. Higher class licensees add new operating privileges, but do not lose any existing privileges when they upgrade. Amateur Extra, Advanced and General class operators may all use these frequencies. (While no new Advanced class licenses will be issued by the FCC, existing licenses will continue to be renewed.) [97.301(d)]

20 METERS

G1A11 What class of amateur license authorizes you to operate on the frequencies 21025 - 21200 kHz and 21300 - 21450 kHz?
A. Amateur Extra class only
B. Amateur Extra and Advanced class only
C. Amateur Extra, Advanced or General class
D. Amateur Extra, Advanced, General or Technician class

C General class operators may operate CW, RTTY and data between 21025 and 21200 kHz. They may also operate CW, phone and image between 21300 and 21450 kHz. Higher class licensees add new operating privileges, but do not lose any existing privileges when they upgrade. Amateur Extra, Advanced and General class operators may all use these frequencies. (While no new Advanced class licenses will be issued by the FCC, existing licenses will continue to be renewed.) [97.301(d)]

15 METERS

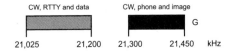

G1B Antenna structure limitations; good engineering and good amateur practice; beacon operation; restricted operation; retransmitting radio signals

G1B01 Up to what height above the ground may you install an antenna structure without needing FCC approval unless your station is in close proximity to an airport as defined in the FCC Rules?
- A. 50 feet
- B. 100 feet
- C. 200 feet
- D. 300 feet

C FCC regulations require approval if your antenna would be more than 200 feet above ground level at its site. This includes the radiating elements, the tower, the supports, and anything else attached to the antenna. (Additional FCC restrictions apply if the antenna is in the vicinity of a public use airport or heliport.) [97.15(a)]

G1B02 If the FCC Rules DO NOT specifically cover a situation, how must you operate your amateur station?
- A. In accordance with general licensee operator principles
- B. In accordance with good engineering and good amateur practice
- C. In accordance with practices adopted by the Institute of Electrical and Electronics Engineers
- D. In accordance with procedures set forth by the International Amateur Radio Union

B It is impossible for rules and regulations to cover every situation that might possibly arise. You are expected to use common sense in those situations where an exact rule does not apply. Your station and its operation should always follow good engineering design and good amateur practice. [97.101(a)]

G1B03 Which of the following types of stations may normally transmit only one-way communications?
- A. Repeater station
- B. Beacon station
- C. HF station
- D. VHF station

B A beacon station normally transmits a signal so that operators may observe propagation and reception characteristics. For this purpose, FCC rules specifically allow an amateur beacon to transmit one-way communications. [97.203(g)]

G1B04 Which of the following does NOT need to be true if an amateur station gathers news information for broadcast purposes?
- A. The information is more quickly transmitted by Amateur Radio
- B. The information must involve the immediate safety of life of individuals or the immediate protection of property
- C. The information must be directly related to the event
- D. The information cannot be transmitted by other means

A   The question is asking which statement does NOT need to be true. The easiest way to answer this is to go through the choices and mark the statements as true or false. Amateurs are not allowed to be involved with any activity related to program production or news gathering for broadcasting to the general public unless it is directly related to an emergency involving an immediate life or property-threatening situation; this makes Answer (B) a true statement. The information to be broadcast must relate directly to the event; this makes (C) a true statement. There must be no other method by which the information can be transmitted; this makes (D) a true statement. The question is asking for the **incorrect** statement; that is Answer (A). It does not matter if the information can be transmitted more quickly by amateur radio. If there is an alternative communication system available, even if it is slower, the amateur station cannot transmit the information for a news broadcast or related program production. [97.113(b)]

G1B05 Under what limited circumstances may music be transmitted by an amateur station?
- A. When it produces no dissonances or spurious emissions
- B. When it is used to jam an illegal transmission
- C. When it is transmitted on frequencies above 1215 MHz
- D. When it is an incidental part of a space shuttle retransmission

D   Normally, music may not be transmitted by an amateur station. This is to avoid infringing upon commercial broadcast activities. Music is often used in communications between ground control and the space shuttle or International Space Station (ISS) for such things as waking up the astronauts in the morning. In this case, since it is an incidental part of the transmission and not the primary purpose of it, the music can be retransmitted by the amateur station along with the rest of the space shuttle transmission, with NASA's permission. [97.113(e)]

G1B06 When may an amateur station in two-way communication transmit a message in a secret code in order to obscure the meaning of the communication?
 A. When transmitting above 450 MHz
 B. During contests
 C. Never
 D. During a declared communications emergency

C An amateur station may never transmit in such a manner as to obscure the meaning of two-way communication. The use of standard abbreviations does not violate this rule, since their meaning is well known. There is one situation that almost seems like an exception: space telecommand operations. When controlling a satellite from a ground station, the transmissions may consist of specially coded messages intended to facilitate communications or related to the function of the spacecraft. Telecommand is not two-way communication, however; it is one-way communication. [97.113(a)(4)]

G1B07 What are the restrictions on the use of abbreviations or procedural signals in the amateur service?
 A. There are no restrictions
 B. They may be used if they do not obscure the meaning of a message
 C. They are not permitted because they obscure the meaning of a message to FCC monitoring stations
 D. Only "10-codes" are permitted

B The use of common abbreviations and procedural signals is standard practice and does not obscure the meaning of a message because their meaning is well known. Any use of abbreviations or codes for the purpose of obscuring the meaning of a communication is prohibited.

There is one exception to this rule: space telecommand operations. When controlling a satellite from a ground station, the transmissions may consist of specially coded messages intended to facilitate communications or related to the function of the spacecraft. FCC Rules permit telecommand stations to transmit special codes intended to obscure the meaning of the telecommand messages to control space stations. (This helps prevent unauthorized stations from transmitting telecommand messages to the spacecraft.) [97.113(a)(4)]

G1B08 When are codes or ciphers permitted in two-way domestic amateur communications?
- A. Never, if intended to obscure meaning
- B. During contests
- C. During nationally declared emergencies
- D. On frequencies above 2.3-GHz

A   The use of codes or ciphers for the purpose of obscuring the meaning of a communication is never allowed. The use of standard abbreviations and procedural signals does not violate this rule since their meaning is commonly known and will not hide the meaning of the message. It does not make any difference whether the communication is domestic or international.

There is one situation that almost seems like an exception to this rule: space telecommand operations. When controlling a satellite from a ground station, the transmissions may consist of specially coded messages intended to facilitate communications or related to the function of the spacecraft. Telecommand is not two-way communication, however, it is one-way communication. [97.113(a)(4)]

G1B09 When are codes or ciphers permitted in two-way international amateur communications?
- A. Never, if intended to obscure meaning
- B. During contests
- C. During internationally declared emergencies
- D. On frequencies above 2.3-GHz

A   The use of codes or ciphers is never allowed if it is intended to obscure the meaning of a communication. The use of standard abbreviations and procedural signals does not violate this rule since their meaning is commonly known and will not hide the meaning of the message. It does not make any difference whether the communication is domestic or international.

There is one situation that almost seems like an exception to this rule: space telecommand operations. When controlling a satellite from a ground station, the transmissions may consist of specially coded messages intended to facilitate communications or related to the function of the spacecraft. Telecommand is not two-way communication, however; it is one-way communication. [97.113(a)(4)]

G1B10 Which of the following amateur transmissions is NOT prohibited by the FCC Rules?
A. The playing of music
B. The use of obscene or indecent words
C. False or deceptive messages or signals
D. Retransmission of space shuttle communications

D   The retransmission of space shuttle (or International Space Station — ISS) communications is allowed provided prior approval for such transmission has been obtained from NASA, and is done for the exclusive use of amateur operators. (NASA has given blanket permission for amateurs to retransmit such communications, provided it is done for educational purposes.)

The playing of music in an amateur transmission is prohibited since it infringes upon commercial broadcast privileges. Common sense, as well as FCC rules, says that using obscene or indecent words or transmitting false or deceptive messages is prohibited. [97.113(a)(4), 97.113(e)]

G1B11 What should you do to keep your station from retransmitting music or signals from a non-amateur station?
A. Turn up the volume of your transceiver
B. Speak closer to the microphone to increase your signal strength
C. Turn down the volume of background audio
D. Adjust your transceiver noise blanker

C   If you are listening to a radio in the background while you are on the air, make sure that the volume of the background radio is turned down low enough that the signal will not be picked up by your microphone. An amateur station is not allowed to transmit music and is not allowed to retransmit signals from a non-amateur station. [97.113(a)(4), 97.113(e)]

G1C Transmitter power standards; certification of external RF-power-amplifiers; standards for certification of external RF-power amplifiers; HF data emission standards

G1C01 What is the maximum transmitting power an amateur station may use on 3690 kHz?
- A. 200 watts PEP output
- B. 1000 watts PEP output
- C. 1500 watts PEP output
- D. The minimum power necessary to carry out the desired communications with a maximum of 2000 watts PEP output

A The general rule is that the maximum power is limited to 1500 watts PEP, although there are exceptions where less power is allowed. The frequency 3690 kHz is in the 80-meter Novice subband, and power for everyone in that area is limited to 200 watts PEP output. [97.313(c)(1)]

To calculate the wavelength, divide 300 by the frequency in megahertz.

$$\text{Wavelength} = \frac{300}{\text{frequency (in MHz)}} = \frac{300}{3.690 \text{ MHz}} = 81 \text{ meters}$$

Although this calculation does not come out exactly as 80 meters, it will help you identify this frequency as being part of the 80-meter band.

G1C02 What is the maximum transmitting power an amateur station may use on 7080 kHz?
- A. 200 watts PEP output
- B. 1000 watts PEP output
- C. 1500 watts PEP output
- D. 2000 watts PEP output

C The general rule is that maximum power is limited to 1500 watts PEP output, although there are exceptions where less power is allowed. The frequency 7080 kHz is below the area where Novices may operate; therefore 1500 watts is allowed. [97.313(a), (b)]

To calculate the wavelength, divide 300 by the frequency in megahertz.

$$\text{Wavelength} = \frac{300}{\text{frequency (in MHz)}} = \frac{300}{7.080 \text{ MHz}} = 42 \text{ meters}$$

Although this calculation does not come out exactly as 40 meters, it will help you identify this frequency as being part of the 40-meter band.

G1C03 What is the maximum transmitting power an amateur station may use on 10.140 MHz?

A. 200 watts PEP output
B. 1000 watts PEP output
C. 1500 watts PEP output
D. 2000 watts PEP output

A The general rule is that maximum power is limited to 1500 watts PEP output, although there are exceptions where less power is allowed. This is the 30-meter band, and the maximum power for everyone in the 30-meter band is 200 watts. (These frequencies are just above the short-wave time broadcasts of WWV and WWVH at 10.0 MHz.) [97.313(c)(1)]

$$\text{Wavelength} = \frac{300}{\text{frequency (in MHz)}} = \frac{300}{10.140 \text{ MHz}} = 30 \text{ meters}$$

This answer will help you identify this frequency as being part of the 30-meter band.

G1C04 What is the maximum transmitting power an amateur station may use on 21.150 MHz?

A. 200 watts PEP output
B. 1000 watts PEP output
C. 1500 watts PEP output
D. 2000 watts PEP output

A The general rule is that the maximum power is limited to 1500 watts PEP output, although there are exceptions where less power is allowed. The frequency 21.150 MHz is in the middle of the 15-meter Novice subband, and the maximum power allowed is 200 watts for everyone. The frequency 21.150 MHz is in the 15-meter band. [97.313(c)(1)]

$$\text{Wavelength} = \frac{300}{\text{frequency (in MHz)}} = \frac{300}{21.105 \text{ MHz}} = 14 \text{ meters}$$

Although this calculation does not come out exactly as 15 meters, it will help you identify this frequency as being part of the 15-meter band.

Commission's Rules

G1C05 What is the maximum transmitting power an amateur station may use on 24.950 MHz?
- A. 200 watts PEP output
- B. 1000 watts PEP output
- C. 1500 watts PEP output
- D. 2000 watts PEP output

C The general rule is that maximum power is limited to 1500 watts PEP output, although there are exceptions where less power is allowed. There are no Novice or Technician operators in the 12-meter band. The maximum power allowed is 1500 watts PEP output. The frequency 24.950 MHz is in the 12-meter band. [97.313(a), (b)]

$$\text{Wavelength} = \frac{300}{\text{frequency (in MHz)}} = \frac{300}{24.955 \text{ MHz}} = 12 \text{ meters}$$

This answer will help you identify this frequency as being part of the 12-meter band.

G1C06 What is the maximum transmitting power an amateur station may use on 3818 kHz?
- A. 200 watts PEP output
- B. 1000 watts PEP output
- C. 1500 watts PEP output
- D. The minimum power necessary to carry out the desired communications with a maximum of 2000 watts PEP output

C The general rule is that maximum power is limited to 1500 watts PEP, although there are exceptions where less power is allowed. The frequency of 3818 kHz is in the 75-meter band. This frequency is not in the Novice subband; therefore, the full 1500 watts PEP output is allowed. [97.313]

$$\text{Wavelength} = \frac{300}{\text{frequency (in MHz)}} = \frac{300}{3.818 \text{ MHz}} = 78 \text{ meters}$$

Although this calculation does not come out exactly as 75 meters, it will help you identify this frequency as being part of the 75-meter band.

G1C07 What is the maximum transmitting power an amateur station may use on 7105 kHz?
- A. 200 watts PEP output
- B. 1000 watts PEP output
- C. 1500 watts PEP output
- D. 2000 watts PEP output

A The general rule is that maximum power is limited to 1500 watts PEP, although there are exceptions where less power is allowed. The frequency of 7105 kHz is in the 40-meter band. This frequency is in the Novice subband, and power for everyone in that area is limited to 200 watts PEP output. [97.313]

$$\text{Wavelength} = \frac{300}{\text{frequency (in MHz)}} = \frac{300}{7.105 \text{ MHz}} = 42 \text{ meters}$$

Although this calculation does not come out exactly as 40 meters, it will help you identify this frequency as being part of the 40-meter band.

G1C08 What is the maximum transmitting power an amateur station may use on 14.300 MHz?
- A. The minimum power necessary to carry out the desired communications with a maximum of 1500 watts PEP output
- B. 200 watts PEP output
- C. 1000 watts PEP output
- D. 2000 watts PEP output

A The general rule is that maximum power is limited to 1500 watts PEP output, although there are exceptions where less power is allowed. The frequency of 14.300 MHz is in the 20-meter band. There is no Novice segment in this band. The maximum power allowed is 1500 watts PEP output. Of course, amateurs are always required to use the mimimum power necessary to carry out the desired communications. [97.313]]

$$\text{Wavelength} = \frac{300}{\text{frequency (in MHz)}} = \frac{300}{14.300 \text{ MHz}} = 21 \text{ meters}$$

Although this calculation does not come out exactly as 20 meters, it will help you identify this frequency as being part of the 20-meter band.

G1C09 What is the absolute maximum transmitting power a General class amateur may use on 28.400 MHz?
- A. 200 watts PEP output
- B. 1000 watts PEP output
- C. 1500 watts PEP output
- D. 2000 watts PEP output

C   The general rule is that maximum power is limited to 1500 watts PEP output, although there are exceptions where less power is allowed. The frequency of 28.400 MHz is in the 10-meter band. Although this is in the Novice segment in this band, General and higher class licensees may use the full maximum power, 1500 watts PEP output. [97.313]

$$\text{Wavelength} = \frac{300}{\text{frequency (in MHz)}} = \frac{300}{28.400 \text{ MHz}} = 10 \text{ meters}$$

This calculation will help you identify this frequency as being part of the 10-meter band.

G1C10 What is the absolute maximum transmitting power a General class amateur may use on 28.150 MHz?
- A. 200 watts PEP output
- B. 1000 watts PEP output
- C. 1500 watts PEP output
- D. 2000 watts PEP output

C   The general rule is that maximum power is limited to 1500 watts PEP output, although there are exceptions where less power is allowed. The frequency of 28.150 MHz is in the 10-meter band. This frequency is in the Novice segment in this band, but General and higher class licensees may use the full maximum power, 1500 watts PEP output. [97.313]

$$\text{Wavelength} = \frac{300}{\text{frequency (in MHz)}} = \frac{300}{28.150 \text{ MHz}} = 10 \text{ meters}$$

This calculation will help you identify this frequency as being part of the 10-meter band.

G1C11 What is the maximum transmitting power an amateur station may use on 1825 kHz?
- A. 200 watts PEP output
- B. 1000 watts PEP output
- C. 2000 watts PEP output
- D. The minimum power necessary to carry out the desired communications with a maximum of 1500 watts PEP output

D The general rule is that maximum power is limited to 1500 watts PEP output, although there are exceptions where less power is allowed. The frequency of 1825 kHz is in the 160-meter band. There is no Novice segment in this band. The maximum power allowed is 1500 watts PEP output. Of course amateurs are always required to use the mimimum power necessary to carry out the desired communications. [97.313]

$$\text{Wavelength} = \frac{300}{\text{frequency (in MHz)}} = \frac{300}{1.825 \text{ MHz}} = 164 \text{ meters}$$

Although this calculation does not come out exactly as 160 meters, it will help you identify this frequency as being part of the 160-meter band.

G1D Examination element preparation; examination administration; temporary station identification

G1D01 What examination elements may you prepare if you hold a General class license?
- A. None
- B. Elements 1 and 2 only
- C. Element 1 only
- D. Elements 1, 2 and 3

B The telegraphy (Morse code) exam element is Element 1. This is the 5 words per minute Morse code exam required for the General and Amateur Extra licenses. Holders of a General class operator license may prepare the telegraphy exam because they already have passed that exam.

Written exam elements are:
Element 2   Technician
Element 3   General
Element 4   Amateur Extra

Holders of a General class operator license may prepare written exam elements for license levels below theirs: Technician. The Technician theory test is Element 2.

General class licensees may prepare the exams for Elements 1 and 2 only. [97.507(a)(2)]

G1D02 What license examinations may you administer if you hold a General class license?
  A. None
  B. General only
  C. Technician and Morse code
  D. Technician, General and Amateur Extra

C   As a General class operator, you may administer license examinations for license levels below yours: Technician, Element 2 and the Morse code exam, Element 1. Of course, you must also be accredited by a Volunteer Examiner Coordinator organization to help administer exams. [97.509(b)(3)(i)]

G1D03 What minimum examination elements must an applicant pass for a Technician license?
  A. Element 2 only
  B. Elements 1 and 2
  C. Elements 2 and 3
  D. Elements 1, 2 and 3

A   The license elements are:

MORSE CODE:
Element 1   5 WPM   General, Amateur Extra
THEORY
Element 2   Technician
Element 3   General
Element 4   Amateur Extra

To earn a Technician class license, an applicant only needs to pass the Element 2 written exam. [97.501(c)]

G1D04 What minimum examination elements must an applicant pass for a Technician license with Morse code credit to operate on the HF bands?
  A. Element 2 only
  B. Elements 1 and 2
  C. Elements 2 and 3
  D. Elements 1, 2 and 3

B   International agreements require that for HF privileges, licensees must demonstrate proficiency in Morse code. For HF privileges, in addition to the Element 2 written exam, a Technician license applicant must pass the 5 wpm Morse code exam (Element 1). [97.301(e), 97.501(c)]

G1D05 What are the requirements for administering Technician examinations?
- A. Three VEC-accredited General class or higher VEs must be present
- B. Two VEC-accredited General class or higher VEs must be present
- C. Two General class or higher VEs must be present, but only one need be VEC accredited
- D. Any two General class or higher VEs must be present

A All license exams are administered through the Volunteer Examiner Coordinator (VEC) system. VEs (Volunteer Examiners) must be accredited by a VEC. There must be three VEC-accredited VEs present at every exam session. Technician exams are administered by General class or higher VEs. [97.509(a), (b)]

G1D06 When may you participate as an administering Volunteer Examiner (VE) for a Technician license examination?
- A. Once you have notified the FCC that you want to give an examination
- B. Once you have a Certificate of Successful Completion of Examination (CSCE) for General class
- C. Once you have prepared telegraphy and written examinations for the Technician license, or obtained them from a qualified supplier
- D. Once you have been granted your FCC General class or higher license and received your VEC accreditation

D All license exams are administered through the Volunteer Examiner Coordinator (VEC) system. VEs (Volunteer Examiners) must be accredited by a VEC. There must be three VEC-accredited VEs present at every exam session. Technician exams are administered by General class or higher VEs. As soon as the FCC issues your General class license (and it appears in the FCC database) and you receive your accreditation from a VEC, you can participate in exam sessions for Technician license exams. [97.509(b)(3)(i)]

G1D07 If you are a Technician licensee with a Certificate of Successful Completion of Examination (CSCE) for General privileges, how do you identify your station when transmitting on 14.035 MHz?

A. You must give your call sign and the location of the VE examination where you obtained the CSCE
B. You must give your call sign, followed by the slant mark "/", followed by the identifier "AG"
C. You may not operate on 14.035 MHz until your new license arrives
D. No special form of identification is needed

B When you qualify for a higher license (in this case, going from Technician to General), you may begin using your new frequency privileges immediately. Until your upgraded license has been granted by the FCC and appears in the FCC database, however, you must identify yourself on the air as a newly upgraded licensee when using your new privileges. Since 14.035 MHz is a General frequency, when transmitting CW you must add the identifier "/AG" after your call sign. [97.119(f)(2)]

$$\text{Wavelength} = \frac{300}{\text{frequency (in MHz)}} = \frac{300}{14.035 \text{ MHz}} = 21 \text{ meters}$$

Although this calculation does not come out exactly as 20 meters, it will help you identify this frequency as being part of the 20-meter band.

G1D08 If you are a Technician licensee with a Certificate of Successful Completion of Examination (CSCE) for General privileges, how do you identify your station when transmitting phone emissions on 14.325 MHz?

A. No special form of identification is needed
B. You may not operate on 14.325 MHz until your new license arrives
C. You must give your call sign, followed by any suitable word that denotes the slant mark and the identifier "AG"
D. You must give your call sign and the location of the VE examination where you obtained the CSCE

C When you qualify for a higher license (in this case, going from Technician to General), you may begin using your new frequency privileges immediately. Until your upgraded license has been granted by the FCC and appears in the FCC database, however, you must identify yourself on the air as a newly upgraded licensee when using your new privileges. Since 14.325 MHz is a General frequency, when transmitting phone you must give your call sign, followed by any suitable word that denotes the slant mark, and then "AG". [97.119(f)(2)]

$$\text{Wavelength} = \frac{300}{\text{frequency (in MHz)}} = \frac{300}{14.325 \text{ MHz}} = 21 \text{ meters}$$

Although this calculation does not come out exactly as 20 meters, it will help you identify this frequency as being part of the 20-meter band.

G1D09 If you are a Technician licensee with a Certificate of Successful Completion of Examination (CSCE) for General privileges, when must you add the special identifier "AG" after your call sign?

A. Whenever you operate using your new frequency privileges
B. Whenever you operate
C. Whenever you operate using Technician frequency privileges
D. A special identifier is not required as long as your General class license application has been filed with the FCC

A When you qualify for a higher license (in this case, going from Technician to General), you may begin using your new frequency privileges immediately. Until your upgraded license has been granted by the FCC and appears in the FCC database, however, you must identify yourself on the air as a newly upgraded licensee anytime you use your new privileges. [97.119(f)(2)]

G1D10 If you are a Technician licensee with a Certificate of Successful Completion of Examination (CSCE) for General privileges, on which of the following band segments must you add the special identifier "AG" after your call sign?
- A. Whenever you operate from 18068 - 18168 kHz
- B. Whenever you operate from 14025 - 14150 kHz and 14225 - 14350 kHz
- C. Whenever you operate from 10100 - 10150 kHz
- D. All of these choices are correct

D When you qualify for a higher license (in this case, going from Technician to General), you may begin using your new frequency privileges immediately. Until your upgraded license has been granted by the FCC and appears in the FCC database, however, you must identify yourself on the air as a newly upgraded licensee anytime you use your new privileges. All of the band segments listed as possible answer choices are General class bands. You must use the special identifier "/AG" when operating on any of them, so all of the choices are correct. [97.119(f)(2)]

G1D11 When may you participate as an administering Volunteer Examiner (VE) to administer the Element 1 5-WPM Morse code examination?
- A. Once you have notified the FCC that you want to give an examination
- B. Once you have a Certificate of Successful Completion of Examination (CSCE) for General class
- C. Once you have prepared telegraphy and written examinations for the Technician license, or obtained them from a qualified supplier
- D. Once you have been granted your FCC General class or higher license and received your VEC accreditation

D All license exams are administered through the Volunteer Examiner Coordinator (VEC) system. VEs (Volunteer Examiners) must be accredited by a VEC. There must be three VEC-accredited VEs present at every exam session. Technician exams are administered by General class or higher VEs. As soon as the FCC issues your General class license (and it appears in the FCC database) and you receive your accreditation from a VEC, you can participate in exam sessions for Technician license exams. [97.509(b)(3)(i)]

G1E Local control; repeater and harmful interference definitions; third party communications

G1E01 As a General class control operator at a Technician station, how must you identify the Technician station when transmitting on 7250 kHz?
- A. With your call sign, followed by the word "controlling" and the Technician call sign
- B. With the Technician call sign, followed by the slant bar "/" (or any suitable word) and your own call sign
- C. With your call sign, followed by the slant bar "/" (or any suitable word) and the Technician call sign
- D. A Technician station should not be operated on 7250 kHz, even with a General control operator

B Both the control operator and the station licensee are responsible for the proper operation of an amateur station. If you are operating a Technician station, the Technician call sign should be included in the identification because that is the station license. Since you, as a General class control operator, are transmitting on a frequency reserved for General class operators, you should give your call sign after the station's call sign. Any operating privileges allowed by your higher (General) class license may be used, as long as the station is properly identified with the Technician call sign followed by the General call sign. Separate the two call signs by using the slant bar (/) or any suitable word used to mean the slant bar. [97.119(e)]

G1E02 Under what circumstances may a 10-meter repeater retransmit the 2-meter signal from a Technician class operator?
- A. Under no circumstances
- B. Only if the station on 10 meters is operating under a Special Temporary Authorization allowing such retransmission
- C. Only during an FCC-declared general state of communications emergency
- D. Only if the 10-meter control operator holds at least a General class license

D FCC rules allow the holder of a Technician, General, Advanced, or Amateur Extra class license to be the control operator of a repeater. The control operator of the repeater must have privileges on the frequency on which the repeater is transmitting, however. A 10-meter repeater must have a General class or higher control operator because lower class licenses don't have privileges on the 10-meter repeater band. A 10-meter repeater may retransmit the 2-meter signal from a Technician class operator because the 10-meter control operator holds at least a General class license. (Of course the operator transmitting to the repeater must have privileges on the frequency on which he or she is transmitting.) [97.205(a)]

G1E03 What kind of amateur station simultaneously retransmits the signals of other stations on a different channel?
  A. Repeater station
  B. Space station
  C. Telecommand station
  D. Relay station

A   A repeater station receives signals on one frequency and retransmits them simultaneously on another frequency. The purpose of a repeater station is to increase the range of portable, mobile, and low-power stations. A repeater is usually located at a high elevation. This gives the signal much more range than it would otherwise have. [97.3(a)(37)]

G1E04 What name is given to a form of interference that seriously degrades, obstructs or repeatedly interrupts a radiocommunication service?
  A. Intentional interference
  B. Harmful interference
  C. Adjacent interference
  D. Disruptive interference

B   Harmful interference can be caused by cable TV systems, computers, power lines, industrial machines, cordless telephones, wireless microphones, and other electrical equipment, as well as from radio communication transmissions. [97.3(a)(22)]

G1E05 What types of messages may be transmitted by an amateur station to a foreign country for a third party?
  A. Messages for which the amateur operator is paid
  B. Messages facilitating the business affairs of any party
  C. Messages of a technical nature or remarks of a personal character
  D. No messages may be transmitted to foreign countries for third parties

C   All communications with a foreign country (including those intended for someone other than the amateur you are talking to — a third party) shall be limited to either remarks of a technical nature (related to radio reception and/or transmission characteristics, for example) or personal remarks that are not important enough to justify using public communication services.

The goal is to avoid competition with the public telecommunication services of other countries. Remember that third-party communications are only allowed with countries that have third-party agreements with the US government. [97.115, 97.117]

G1E06 If a repeater is causing harmful interference to another repeater and a frequency coordinator has recommended the operation of one station only, who is responsible for resolving the interference?
- A. The licensee of the unrecommended repeater
- B. Both repeater licensees
- C. The licensee of the recommended repeater
- D. The frequency coordinator

A A recommended (coordinated) repeater takes precedence over an unrecommended (uncoordinated) repeater. The licensee of the unrecommended repeater is primarily responsible for resolving the interference. [97.205(c)]

G1E07 If a repeater is causing harmful interference to another amateur repeater and a frequency coordinator has recommended the operation of both stations, who is responsible for resolving the interference?
- A. The licensee of the repeater that has been recommended for the longest period of time
- B. The licensee of the repeater that has been recommended the most recently
- C. The frequency coordinator
- D. Both repeater licensees

D If both repeaters are recommended (coordinated), they have equal responsibility for resolving the interference. [97.205(c)]

G1E08 If a repeater is causing harmful interference to another repeater and a frequency coordinator has NOT recommended either station, who is primarily responsible for resolving the interference?
- A. Both repeater licensees
- B. The licensee of the repeater that has been in operation for the longest period of time
- C. The licensee of the repeater that has been in operation for the shortest period of time
- D. The frequency coordinator

A If neither repeater station is recommended (coordinated), they both have equal responsibility for resolving the interference. [97.205(c)]

G1E09 If the FCC rules say that the amateur service is a secondary user of a frequency band, and another service is a primary user, what does this mean?

    A. Nothing special; all users of a frequency band have equal rights to operate

    B. Amateurs are only allowed to use the frequency band during emergencies

    C. Amateurs are allowed to use the frequency band only if they do not cause harmful interference to primary users

    D. Amateurs must increase transmitter power to overcome any interference caused by primary users

C   You should always listen before you transmit. This is especially important on bands where the Amateur Service is a secondary user. Amateurs are only permitted to use these frequencies if they do not cause harmful interference to the primary users. [97.303]

G1E10 If you are using a frequency within a band assigned to the amateur service on a secondary basis, and a station assigned to the primary service on that band causes interference, what action should you take?

    A. Notify the FCC's regional Engineer in Charge of the interference

    B. Increase your transmitter's power to overcome the interference

    C. Attempt to contact the station and request that it stop the interference

    D. Change frequencies; you may be causing harmful interference to the other station, in violation of FCC rules

D   You should always listen before you transmit. This is especially important on bands where the Amateur Service is a secondary user. Amateurs are only permitted to use these frequencies if they do not cause harmful interference to the primary users. If you hear a station in the primary service or receive interference from such a station you should immediately change frequencies. Otherwise you might be causing interference to the primary station. [97.303]

G1E11 If you are using a language besides English to make a contact, what language must you use when identifying your station?
- A. The language being used for the contact
- B. The language being used for the contact, provided the US has a third-party communications agreement with that country
- C. English
- D. Any language of a country that is a member of the International Telecommunication Union

C   The FCC requires you to give your station identification in English. You can use any language to communicate with another amateur. Practice your German with someone from Germany or your Japanese with another ham in Japan, but be sure to give your station identification in English every 10 minutes and at the end of your contact. [97.119(b)(2)]

G1F Certification of external RF-power-amplifiers; standards for certification of external RF-power amplifiers; HF data emission standards

G1F01 External RF power amplifiers designed to operate below what frequency may require FCC certification?
- A. 28 MHz
- B. 35 MHz
- C. 50 MHz
- D. 144 MHz

D   144 MHz is the bottom of the 2-meter band. Frequencies below 144 MHz have relatively long-range signal transmission. There has been a specific problem with the use of amplifiers on the Citizens' Band (CB) Radio Service, which is near the amateur 10-meter band. FCC Rules are designed to prevent this illegal operation. FCC Certification means that the manufacturer has submitted the equipment specifications or an actual unit to the FCC so they can assure the equipment will meet all FCC requirements. [97.315(a)]

$$\text{Wavelength} = \frac{300}{\text{frequency (in MHz)}} = \frac{300}{144 \text{ MHz}} = 2 \text{ meters}$$

This calculation will help you identify this frequency as being the low-frequency end of the 2-meter band.

G1F02 Without a grant of FCC certification, how many external RF amplifiers of a given design capable of operation below 144 MHz may you build or modify in one calendar year?
   A. None
   B. 1
   C. 5
   D. 10

B   An amateur may modify or build an amplifier capable of operating below 144 MHz, but only one per year of a specific type. This is, in part, to ensure that the amateur operator is not commercially producing and/or modifying the amplifiers. 144 MHz is the bottom of the 2-meter band. Frequencies below 144 MHz have relatively long-range signal transmission. There has been a specific problem with the use of amplifiers on the Citizens' Band (CB) Radio Service, which is near the amateur 10-meter band. FCC Rules are designed to prevent this illegal operation. [97.315(a)]

$$\text{Wavelength} = \frac{300}{\text{frequency (in MHz)}} = \frac{300}{144 \text{ MHz}} = 2 \text{ meters}$$

This calculation will help you identify this frequency as being the low-frequency end of the 2-meter band.

G1F03 Which of the following standards must be met if FCC certification of an external RF amplifier is required?
   A. The amplifier must not be able to amplify a 28-MHz signal to more than ten times the input power
   B. The amplifier must not be capable of reaching its designed output power when driven with less than 50 watts
   C. The amplifier must not be able to be operated for more than ten minutes without a time delay circuit
   D. The amplifier must not be able to be modified by an amateur operator

B   An amplifier must not be capable of reaching the designed output power with less than 50 W of drive power. The FCC is attempting to make sure that an amplifier intended for the amateur service is not used in another service, such as the Citizens' Band (CB) Radio Service. [97.317(a)(3)]

G1F04 Which of the following would NOT disqualify an external RF power amplifier from being granted FCC certification?
- A. The capability of being modified by the operator for use outside the amateur bands
- B. The capability of achieving full output power when driven with less than 50 watts
- C. The capability of achieving full output power on amateur frequencies between 24 and 35 MHz
- D. The capability of being switched by the operator to all amateur frequencies below 24 MHz

D The easiest way to determine the correct answer is to go through the choices and mark the statements as true or false. The one statement marked false will be the correct answer. To prevent operation of external amplifiers outside the amateur service, specifically in the CB Radio Service, the FCC has certain standards for Certification. If the amplifier can be modified for use outside the amateur bands, this will disqualify it; (A) is a true statement. (Notice that this requirement does not prohibit an amateur from modifying an amplifier to operate on amateur bands.)

One of the requirements is that the amplifier not incorporate more gain than necessary for amateur use, so it must not be capable of achieving the designed power output when driven with less than 50 watts. If it can achieve full power output when driven with less than 50 watts, this would disqualify the amplifier; (B) is a true statement.

To gain certification, the amplifier must not be able to be operated between 24 MHz and 35 MHz. If it can, this will disqualify it from certification; (C) is a true statement.

Manufacturers may build amplifiers designed to operate on amateur frequencies below 24 MHz, and this often involves some type of band switch circuits. The manufacturers may receive FCC Certification for those amplifiers. Statement (D) would not disqualify the amplifier from Certification, so that is the correct answer. [97.317(b), (c)]

G1F05 What is the maximum symbol rate permitted for packet emissions below 28 MHz?
- A. 300 bauds
- B. 1200 bauds
- C. 19.6 kilobauds
- D. 56 kilobauds

A For frequencies below 28 MHz, the maximum symbol rate for packet radio emissions is 300 bauds. 28 MHz is the bottom, or low-frequency end of the 10-meter band. The lower the frequency, the slower the allowable sending rate for packet (data) or RTTY emissions. This is because the amateur bands below 28 MHz are relatively narrow. The faster the transmission speed, the wider the signal bandwidth needs to be to convey the information. By limiting the allowable sending rate, the bandwidth occupied by the transmission is also limited. [97.305(c), 97.307(f)(3)]

G1F06 What is the maximum symbol rate permitted for RTTY emissions below 28 MHz?
- A. 56 kilobauds
- B. 19.6 kilobauds
- C. 1200 bauds
- D. 300 bauds

D For frequencies below 28 MHz, the maximum symbol rate for RTTY emissions is 300 bauds. 28 MHz is the bottom, or low-frequency end of the 10-meter band. The lower the frequency, the slower the allowable sending rate for packet (data) or RTTY emissions. This is because the amateur bands below 28 MHz are relatively narrow. The faster the transmission speed, the wider the signal bandwidth needs to be to convey the information. By limiting the allowable sending rate, the bandwidth occupied by the transmission is also limited. [97.305(c), 97.307(f)(3)]

G1F07 What is the maximum symbol rate permitted for packet emissions on the 10-meter band?
- A. 300 bauds
- B. 1200 bauds
- C. 19.6 kilobauds
- D. 56 kilobauds

B   The 10-meter band starts at 28 MHz. 28 MHz marks a change point in the FCC rules for allowable sending speeds, and above this frequency you can use a 1200 baud symbol rate. The lower the frequency, the slower the allowable sending rate for packet (data) or RTTY emissions. This is because the amateur bands below 28 MHz are relatively narrow, and they are wider at higher frequencies. The faster the transmission speed, the wider the signal bandwidth needs to be to convey the information. By limiting the allowable sending rate, the bandwidth occupied by the transmission is also limited. [97.307(f)(4)]

G1F08 What is the maximum symbol rate permitted for packet emissions on the 2-meter band?
- A. 300 bauds
- B. 1200 bauds
- C. 19.6 kilobauds
- D. 56 kilobauds

C   The 2-meter band is from 144 MHz to 148 MHz. For frequencies above 50 MHz, the FCC rules allow a maximum symbol rate of 19.6 kilobauds for data (such as packet radio) emissions. [97.307(f)(5)]

G1F09 What is the maximum symbol rate permitted for RTTY or data emissions on the 10-meter band?
- A. 56 kilobauds
- B. 19.6 kilobauds
- C. 1200 bauds
- D. 300 bauds

C   The 10-meter band starts at 28 MHz. 28 MHz marks a change point in the FCC rules for allowable sending speeds, and above this frequency you can use a 1200 baud symbol rate for RTTY or data emissions. The lower the frequency, the slower the allowable sending rate for packet (data) or RTTY emissions. This is because the amateur bands below 28 MHz are relatively narrow, and they are wider at higher frequencies. The faster the transmission speed, the wider the signal bandwidth needs to be to convey the information. By limiting the allowable sending rate, the bandwidth occupied by the transmission is also limited. [97.307(f)(4)]

G1F10 What is the maximum symbol rate permitted for RTTY or data emissions on the 6- and 2-meter bands?
 A. 56 kilobauds
 B. 19.6 kilobauds
 C. 1200 bauds
 D. 300 bauds

B The 6-meter band is from 50 MHz to 54 MHz. For frequencies above 50 MHz, the FCC Rules allow a maximum symbol rate of 19.6 kilobauds for RTTY and data emissions. [97.307(f)(5)]

G1F11 What is the maximum authorized bandwidth of RTTY, data or multiplexed emissions using an unspecified digital code on the 6- and 2-meter bands?
 A. 20 kHz
 B. 50 kHz
 C. The total bandwidth shall not exceed that of a single-sideband phone emission
 D. The total bandwidth shall not exceed 10 times that of a CW emission

A On the 6 and 2-meter bands, the FCC rules allow a maximum bandwidth of 20 kHz for RTTY, data and multiplexed emissions using an unspecified digital code. [97.307(f)(5)]

Subelement G2

Operating Procedures

There will be 6 questions on your general class exam from the Operating Procedures subelement. Those 6 questions will be taken from the 6 groups of questions labeled G2A through G2F, printed in this chapter.

G2A Phone operating procedures

G2A01 Which sideband is commonly used for 20-meter phone operation?
   A. Upper
   B. Lower
   C. Amplitude compandored
   D. Double

A Amateurs normally use the upper sideband for 20-meter phone operation. By suppressing the carrier and one sideband, the bandwidth required for a phone transmission is greatly reduced, and more operators can be accommodated on the band. Whether the upper or lower sideband is used is strictly a matter of convention, and not of regulation. The convention to use the lower sideband on the 160-meter, 75/80-meter, and 40-meter bands and the upper sideband on the higher-frequency (shorter-wavelength) bands developed from the design requirements of early SSB transmitters. Although modern amateur SSB equipment is more flexible, the convention persists. If everyone else in a particular band is using a particular sideband, you will need to use the same one in order to be able to communicate.

G2A02 Which sideband is commonly used on 3925 kHz for phone operation?
   A. Upper
   B. Lower
   C. Amplitude compandored
   D. Double

B   The frequency 3925 kHz is in the 75-meter phone band.

$$\text{Wavelength} = \frac{300}{\text{frequency (in MHz)}} = \frac{300}{3.925 \text{ MHz}} = 76 \text{ meters}$$

(Although this calculation does not come out exactly as 75 meters, it will help you identify this frequency as being part of the 75-meter band.)

Amateurs normally use the lower sideband for 75/80-meter phone operation. By suppressing the carrier and one sideband, the bandwidth required for a phone transmission is greatly reduced, and more operators can be accommodated on the band. Whether the upper or lower sideband is used is strictly a matter of convention, and not of regulation. The convention to use the lower sideband on the 160-meter, 75/80-meter, and 40-meter bands and the upper sideband on the higher-frequency (shorter-wavelength) bands developed from the design requirements of early SSB transmitters. Although modern amateur SSB equipment is more flexible, the convention persists. If everyone else in a particular band is using a particular sideband, you will need to use the same one in order to be able to communicate.

G2A03 Which sideband is commonly used for 40-meter phone operation?
   A. Upper
   B. Lower
   C. Amplitude compandored
   D. Double

B   By suppressing the carrier and one sideband, the bandwidth required for a phone transmission is greatly reduced, and more operators can be accommodated on the band. Whether the upper or lower sideband is used is strictly a matter of convention, and not of regulation. The convention to use the lower sideband on the 160, 80, and 40-meter bands and the upper sideband on the higher-frequency (shorter-wavelength) bands developed from the design requirements of early SSB transmitters. Although modern amateur SSB equipment is more flexible, the convention persists. If everyone else in a particular band is using a particular sideband, you will need to use the same one in order to be able to communicate.

G2A04 Which sideband is commonly used for 10-meter phone operation?

    A. Double
    B. Lower
    C. Amplitude compandored
    D. Upper

D   By suppressing the carrier and one sideband, the bandwidth required for a phone transmission is greatly reduced, and more operators can be accommodated on the band. Whether the upper or lower sideband is used is strictly a matter of convention, and not of regulation. The convention to use the lower sideband on the 160, 80, and 40-meter bands and the upper sideband on the 20, 17, 15, 12 and 10-meter bands developed from the design requirements of early SSB transmitters. Although modern amateur SSB equipment is more flexible, the convention persists. If everyone else in a particular band is using a particular sideband, you will need to use the same one in order to be able to communicate.

G2A05 Which sideband is commonly used for 15-meter phone operation?

    A. Upper
    B. Lower
    C. Amplitude compandored
    D. Double

A   By suppressing the carrier and one sideband, the bandwidth required for a phone transmission is greatly reduced, and more operators can be accommodated on the band. Whether the upper or lower sideband is used is strictly a matter of convention, and not of regulation. The convention to use the lower sideband on the 160, 80, and 40-meter bands and the upper sideband on the 20, 17, 15, 12 and 10-meter bands developed from the design requirements of early SSB transmitters. Although modern amateur SSB equipment is more flexible, the convention persists. If everyone else in a particular band is using a particular sideband, you will need to use the same one in order to be able to communicate.

G2A06 Which sideband is commonly used for 17-meter phone operation?
- A. Lower
- B. Upper
- C. Amplitude compandored
- D. Double

B By suppressing the carrier and one sideband, the bandwidth required for a phone transmission is greatly reduced, and more operators can be accommodated on the band. Whether the upper or lower sideband is used is strictly a matter of convention, and not of regulation. The convention to use the lower sideband on the 160, 80, and 40-meter bands and the upper sideband on the 20, 17, 15, 12 and 10-meter bands developed from the design requirements of early SSB transmitters. Although modern amateur SSB equipment is more flexible, the convention persists. If everyone else in a particular band is using a particular sideband, you will need to use the same one in order to be able to communicate.

G2A07 Which of the following modes of voice communication is most commonly used on the High Frequency Amateur bands?
- A. Frequency modulation (FM)
- B. Amplitude modulation (AM)
- C. Single sideband (SSB)
- D. Phase modulation (PM)

C Most amateurs who use voice communications on the high frequency bands use single sideband (SSB) voice. There are some operators who prefer the high-fidelity audio of double-sideband full-carrier amplitude modulation (AM). AM requires more than twice the bandwidth of an SSB signal, however. There is also some frequency modulation (FM) and phase modulation (PM) voice operation on the 10-meter band, but that mode also requires a much wider bandwidth.

G2A08 Why is the single sideband mode of voice transmission used more frequently than Amplitude Modulation (AM) on the HF amateur bands?
- A. Single sideband transmissions use less spectrum space
- B. Single sideband transmissions are more power efficient
- C. No carrier is transmitted with a single sideband transmission
- D. All of the above responses are correct

D Single sideband (SSB) voice communication is used much more frequently than double-sideband, full-carrier amplitude modulation (AM) on the HF bands because it uses less than half as much spectrum space. The RF carrier is not transmitted with an SSB signal. That means SSB transmissions are more power efficient, since the full transmitter power can be used to transmit the one sideband rather than being divided between the two sidebands and the carrier.

G2A09 Which of the following statements is true of a lower sideband transmission?

A. It is called lower sideband because the lower sideband is greatly attenuated
B. It is called lower sideband because the lower sideband is the only sideband transmitted, since the upper sideband is suppressed
C. The lower sideband is wider than the upper sideband
D. The lower sideband is the only sideband that is authorized on the 160-, 75- and 40-meter amateur bands

B Single sideband (SSB) voice transmissions are identified by which sideband is used. If the sideband with a frequency lower than the RF carrier frequency is used, then the signal is known as a lower sideband (LSB) transmission. If the sideband with a frequency higher than the RF carrier frequency is used, then the signal is known as an upper sideband (USB) transmission. In both cases the opposite sideband is suppressed. Amateurs normally use lower sideband on bands with frequencies less than 7.3 MHz (the 160, 75/80 and 40-meter bands), and upper sideband on bands with frequencies higher than 14 MHz (the 20, 17, 15, 12 and 10-meter bands). This is not a requirement of the FCC Rules in Part 97, though. It is simply by common agreement.

G2A10 Which of the following statements is true of an upper sideband transmission?

A. Only the upper sideband is transmitted, since the opposite sideband is suppressed
B. The upper sideband is greatly attenuated as compared with the carrier
C. The upper sideband is greatly attenuated as compared with the lower sideband
D. Only the upper sideband may be used for phone transmissions on the amateur bands with frequencies above 14 MHz

A Single sideband (SSB) voice transmissions are identified by which sideband is used. If the sideband with a frequency lower than the RF carrier frequency is used, then the signal is known as a lower sideband (LSB) transmission. If the sideband with a frequency higher than the RF carrier frequency is used, then the signal is known as an upper sideband (USB) transmission. In both cases the opposite sideband is suppressed. Amateurs normally use lower sideband on bands with frequencies less than 7.3 MHz (the 160, 75/80 and 40-meter bands), and upper sideband on bands with frequencies higher than 14 MHz (the 20, 17, 15, 12 and 10-meter bands). This is not a requirement of the FCC Rules in Part 97, though. It is simply by common agreement.

G2A11 Why do most amateur stations use lower sideband on the 160-, 75- and 40-meter bands?
  A. The lower sideband is more efficient at these frequency bands
  B. The lower sideband is the only sideband legal on these frequency bands
  C. Because it is fully compatible with an AM detector
  D. Current amateur practice is to use lower sideband on these frequency bands

D There are no FCC Part 97 Rules about which sideband is used on any band. Amateurs have adopted the practice of using lower sideband on bands with frequencies less than 7.3 MHz (the 160, 75/80 and 40-meter bands), and upper sideband on bands with frequencies higher than 14 MHz (the 20, 17, 15, 12 and 10-meter bands).

G2B Operating courtesy

G2B01 If you are the net control station of a daily HF net, what should you do if the frequency on which you normally meet is in use just before the net begins?
  A. Reduce your output power and start the net as usual
  B. Increase your power output so that net participants will be able to hear you over the existing activity
  C. Cancel the net for that day
  D. Conduct the net on a clear frequency 3 to 5-kHz away from the regular net frequency

D Operating courtesy and common sense require that if the frequency is already in use, you should move to another frequency to conduct the net. For single-sideband phone (SSB), this means you need to move from 3 to 5 kHz away in order to avoid interference.

G2B02 If a net is about to begin on a frequency which you and another station are using, what should you do?
  A. As a courtesy to the net, move to a different frequency
  B. Increase your power output to ensure that all net participants can hear you
  C. Transmit as long as possible on the frequency so that no other stations may use it
  D. Turn off your radio

A It is much easier for you and one other station to move to a different frequency than it would be for the entire net to move to a different frequency. If you do not move to a different frequency to complete your QSO (conversation), the net will move instead, but this may cause considerable inconvenience and confusion to the net participants.

G2B03 If propagation changes during your contact and you notice increasing interference from other activity on the same frequency, what should you do?
- A. Tell the interfering stations to change frequency, since you were there first
- B. Report the interference to your local Amateur Auxiliary Coordinator
- C. Turn on your amplifier to overcome the interference
- D. Move your contact to another frequency

D Operating courtesy and common sense suggest that whoever can most easily resolve an interference problem be the one to do so. If you begin to have interference from other activity on the same frequency, moving your contact to another frequency is the courteous thing to do.

G2B04 When selecting a CW transmitting frequency, what minimum frequency separation from a contact in progress should you allow to minimize interference?
- A. 5 to 50 Hz
- B. 150 to 500 Hz
- C. 1 to 3 kHz
- D. 3 to 6 kHz

B The more bandwidth a mode (emission type) takes up, the more frequency separation you will need from a contact currently in progress to avoid interference. CW emissions require the least bandwidth and need the least frequency separation. Most radios use narrow filters for CW reception, so you should be able to select an operating frequency within about 150 to 500 Hz of another CW station.

G2B05 When selecting a single-sideband phone transmitting frequency, what minimum frequency separation from a contact in progress should you allow (between suppressed carriers) to minimize interference?
- A. 150 to 500 Hz
- B. Approximately 3 kHz
- C. Approximately 6 kHz
- D. Approximately 10 kHz

B The more bandwidth a mode (emission type) takes up, the more frequency separation you will need from a contact currently in progress to avoid interference. Single-sideband phone (SSB) requires considerably more bandwidth than CW, and therefore much more frequency separation between contacts. You will need approximately 3-kHz of separation from another contact in progress.

G2B06 When selecting a RTTY transmitting frequency, what minimum frequency separation from a contact in progress should you allow (center to center) to minimize interference?
   A. 60 Hz
   B. 250 to 500 Hz
   C. Approximately 3 kHz
   D. Approximately 6 kHz

B   The more bandwidth a mode (emission type) takes up, the more frequency separation you will need from a contact currently in progress to avoid interference. RTTY requires slightly more bandwidth than CW emissions, and therefore slightly more frequency separation between contacts. You should tune to a frequency about 250 to 500 Hz away from a RTTY contact in progress.

G2B07 What is a band plan?
   A. A voluntary guideline beyond the divisions established by the FCC for using different operating modes within an amateur band
   B. A guideline from the FCC for making amateur frequency band allocations
   C. A plan of operating schedules within an amateur band published by the FCC
   D. A plan devised by a club to best use a frequency band during a contest

A   The goal of a band plan is to minimize interference between the various modes that share each band. This is done by setting aside certain sections of a band for each different operating mode, such as FM repeaters, simplex activity, CW, SSB, AM, satellite, radio control operations, etc. Band plans are voluntary and are agreed to nationally and/or internationally. There are band plans for each IARU (International Amateur Radio Union) region, as well as national band plans like the ARRL's band plans.

ARRL Band Plan Summary for HF Bands

Frequency	Mode
160 Meters (1.8-2.0 MHz):	
1.800-1.810	Data
1.810	CW QRP
1.800-2.000	CW
1.843-2.000	SSB, SSTV and other wideband modes
1.910	SSB QRP
1.995-2.000	Experimental
1.999-2.000	Beacons
80 Meters (3.5-4.0 MHz):	
3.590	Data DX
3.580-3.620	Data
3.620-3.635	Packet
3.635-3.750	CW
3.790-3.800	Phone DX window
3.845	SSTV
3.885	AM calling frequency
40 Meters (7.0-7.3 MHz):	
7.040	Data DX/QRP calling
7.080-7.100	Data
7.171	SSTV
7.290	AM calling frequency
30 Meters (10.1-10.15 MHz):	
10.130-10.140	Data
10.140-10.150	Packet
20 Meters (14.0-14.35 MHz):	
14.070-14.095	Data
14.095-14.0995	Packet
14.100	NCDXF/IARU Beacons
14.1005-14.112	Packet/SSB
14.230	SSTV
14.286	AM calling frequency
17 Meters (18.068-18.168 MHz):	
18.100-18.105	Data
18.105-18.110	Packet
15 Meters (21.0-21.45 MHz):	
21.070-21.090	Data
21.090-21.125	Packet
21.340	SSTV
12 Meters (24.89-24.99 MHz):	
24.920 24.925	Data
24.925 24.930	Packet
10 Meters (28-29.7 MHz):	
28.000-28.070	CW
28.070-28.120	Data
28.120-28.189	Packet
28.150-28.190	CW
28.200-28.300	Beacons
28.300-29.300	Phone
28.680	SSTV
29.000-29.200	AM
29.300-29.510	Satellite Downlinks
29.520-29.590	Repeater Inputs
29.600	FM Simplex
29.610-29.700	Repeater Outputs

G2B08 What is another name for a voluntary guideline that guides amateur activities and extends beyond the divisions established by the FCC for using different operating modes within an amateur band?
   A. A "Band Plan"
   B. A "Frequency and Solar Cycle Guide"
   C. The "Knowledgeable Operator's Guide"
   D. The "Frequency Use Guidebook"

A The goal of a band plan is to minimize interference between the various modes that share each band. This is done by setting aside certain sections of a band for each different operating mode, such as FM repeaters, simplex activity, CW, SSB, AM, satellite, radio control operations, etc. Band plans are voluntary and are agreed to nationally and/or internationally. Band plans are more specific guidelines for each mode than the FCC Part 97 Rules.

G2B09 When choosing a frequency for Slow-Scan TV (SSTV) operation, what should you do to comply with good amateur practice?
   A. Review FCC Part 97 Rules regarding permitted frequencies and emissions
   B. Follow generally accepted gentlemen's agreement band plans
   C. Before transmitting, listen to the frequency to be used to avoid interfering with an ongoing communication
   D. All of these choices

D When you want to choose an operating frequency for any mode (such as for SSTV operation) you should review the FCC Part 97 Rules to be sure you know the frequencies where that mode is permitted. You should also follow the specific band plan agreements for that mode. Finally, always listen before you transmit, to avoid interference to other communications.

ARRL Band Plan for SSTV Frequencies

Band (Meters)	SSTV Frequencies (MHz)
160	1.843 - 2.00
75	3.845
40	7.171
20	14.230
15	21.340
10	28.680

G2B10 When choosing a frequency for radioteletype (RTTY) operation, what should you do to comply with good amateur practice?
- A. Review FCC Part 97 Rules regarding permitted frequencies and emissions
- B. Follow generally accepted gentlemen's agreement band plans
- C. Before transmitting, listen to the frequency to be used to avoid interfering with an ongoing communication
- D. All of these choices are correct

D When you want to choose an operating frequency for any mode (such as for RTTY operation) you should review the FCC Part 97 Rules to be sure you know the frequencies where that mode is permitted. You should also follow the specific band plan agreements for that mode. Finally, always listen before you transmit, to avoid interference to other communications.

ARRL Band Plan for RTTY/Data Frequencies

Band (Meters)	RTTY/Data Frequencies (MHz)
160	1.800 - 1.810
80	3.580 - 3.620
40	7.080 - 7.100
30	10.130 - 10.140
20	14.070 - 14.095
17	18.100 - 18.105
15	21.070 - 21.090
12	24.920 - 24.925
10	28.070 - 28.120

G2B11 When choosing a frequency for HF Packet operation, what should you do to comply with good amateur practice?
  A. Review FCC Part 97 Rules regarding permitted frequencies and emissions
  B. Follow generally accepted gentlemen's agreement band plans
  C. Before transmitting, first listen on the frequency to be used to avoid interfering with an ongoing communication
  D. All of these choices

D   When you want to choose an operating frequency for any mode (such as for HF packet operation) you should review the FCC Part 97 Rules to be sure you know the frequencies where that mode is permitted. You should also follow the specific band plan agreements for that mode. Finally, always listen before you transmit, to avoid interference to other communications.

ARRL Band Plan for Packet Frequencies

Band (Meters)	Packet Frequencies (MHz)
160	1.800 - 1.810
80	3.620 - 3.635
30	10.140 - 10.150
20	14.095 - 14.0995
20	14.1005 - 14.112
17	18.105 - 18.110
15	21.090 - 21.125
12	24.925 - 24.930
10	28.120 - 28.189

G2C Emergencies, including drills and emergency communications

G2C01 What means may an amateur station in distress use to attract attention, make known its condition and location, and obtain assistance?
    A. Only Morse code signals sent on internationally recognized emergency channels
    B. Any means of radiocommunication, but only on internationally recognized emergency channels
    C. Any means of radiocommunication
    D. Only those means of radiocommunication for which the station is licensed

C   When normal communications are not available and the immediate safety of human life or protection of property is involved, all of the normal rules for an amateur station may be broken in order to obtain assistance. This means that any method of communication, on any frequency, and at any power output, may be used to communicate and resolve the emergency. Just be sure you have a real emergency!

G2C02 During a disaster in the US, when may an amateur station make transmissions necessary to meet essential communication needs and assist relief operations?
    A. When normal communication systems are overloaded, damaged or disrupted
    B. Only when the local RACES net is activated
    C. Never; only official emergency stations may transmit in a disaster
    D. When normal communication systems are working but are not convenient

A   The normal communication system must be damaged or disrupted or overloaded in order for stations to become involved in relief operations. It is not enough for normal communication systems to simply be inconvenient, to justify using amateur transmissions instead. A local RACES (Radio Amateur Civil Emergency Service) net is authorized by the local civil defense agency. The FCC authorizes RACES operation after local, state or federal officials have requested it. It is limited to official civil defense activities. In the event of a disaster, it is not necessary for the local RACES net to be activated for amateurs to meet essential communications needs.

G2C03 If a disaster disrupts normal communications in your area, what may the FCC do?
- A. Declare a temporary state of communication emergency
- B. Temporarily seize your equipment for use in disaster communications
- C. Order all stations across the country to stop transmitting at once
- D. Nothing until the President declares the area a disaster area

A   If the FCC declares a temporary state of communication emergency, the FCC engineer in charge may designate certain frequencies for use only by stations assisting the stricken area. Amateur stations may also be designated to police the emergency frequencies, warning noncomplying stations that may be operating there. Other special conditions and rules may be set forth that will apply during the communications emergency.

G2C04 If a disaster disrupts normal communications in an area what would the FCC include in any notice of a temporary state of communication emergency?
- A. Any additional test questions needed for the licensing of amateur emergency communications workers
- B. A list of organizations authorized to temporarily seize your equipment for disaster communications
- C. Any special conditions requiring the use of non-commercial power systems
- D. Any special conditions and special rules to be observed by stations during the emergency

D   If the FCC declares a temporary state of communication emergency, the FCC engineer in charge may designate certain frequencies for use only by stations assisting the stricken area. Amateur stations may also be designated to police the emergency frequencies, warning noncomplying stations that may be operating there. The FCC notice would include any other special conditions and rules to be observed by stations during the communications emergency.

G2C05 During an emergency, what power output limitations must be observed by a station in distress?
   A. 200 watts PEP
   B. 1500 watts PEP
   C. 1000 watts PEP during daylight hours, reduced to 200 watts PEP during the night
   D. There are no limitations during an emergency

D   When normal communications are not available and the immediate safety of human life or protection of property is involved, all of the normal rules for an amateur station may be broken in order to obtain assistance. This means that any method of communication, on any frequency, and at any power output, may be used to communicate and resolve the emergency. Just be sure you have a real emergency!

G2C06 During a disaster in the US, what frequencies may be used to obtain assistance?
   A. Only frequencies in the 80-meter band
   B. Only frequencies in the 40-meter band
   C. Any frequency
   D. Any United Nations approved frequency

C   When normal communications are not available and the immediate safety of human life or protection of property is involved, all of the normal rules for an amateur station may be broken in order to obtain assistance. This means that any method of communication, on any frequency, and at any power output, may be used to communicate and resolve the emergency. It doesn't matter if the distress is personal to the station, or a general disaster. Just be sure you have a real emergency!

G2C07 If you are communicating with another amateur station and hear a station in distress break in, what should you do?
- A. Continue your communication because you were on frequency first
- B. Acknowledge the station in distress and determine its location and what assistance may be needed
- C. Change to a different frequency so the station in distress may have a clear channel to call for assistance
- D. Immediately cease all transmissions because stations in distress have emergency rights to the frequency

B Whenever you hear a station in distress (where there is immediate threat to human life or property), you should take whatever action is necessary to determine what assistance that station needs and attempt to provide it. Don't assume that some other station will handle the emergency; you may be the only station receiving the one in distress.

G2C08 Why do stations in the Radio Amateur Civil Emergency Service (RACES) participate in training tests and drills?
- A. To practice orderly and efficient operations for the civil defense organization they serve
- B. To ensure that members attend monthly on-the-air meetings
- C. To ensure that RACES members are able to conduct tests and drills
- D. To acquaint members of RACES with other members they may meet in an emergency

A RACES (Radio Amateur Civil Emergency Service) stations are authorized by the local civil defense organization. Tests and drills are important to ensure that communications are handled in an orderly and efficient manner for the civil defense organization in the event of a true emergency.

G2C09 When are you prohibited from helping a station in distress?
- A. When that station is not transmitting on amateur frequencies
- B. When the station in distress offers no call sign
- C. You are not ever prohibited from helping any station in distress
- D. When the station is not another amateur station

C Whenever you hear a station in distress (where there is immediate threat to human life or property), you should take whatever action is necessary to determine what assistance that station needs and attempt to provide it. Don't assume that some other station will handle the emergency; you may be the only station receiving the one in distress. This means that any method of communication, on any frequency, and at any power output, may be used to communicate and resolve the emergency.

G2C10 When FCC declares a temporary state of communication emergency, what must you do?
  A. Abide by the limitations or conditions set forth in the FCC notice
  B. Stay off the air until 30 days after FCC lifts the emergency notice
  C. Only communicate with stations within 2 miles of your location
  D. Nothing; wait until the President declares a formal emergency before taking further action

A   If the FCC declares a temporary state of communication emergency, the FCC engineer in charge may designate certain frequencies for use only by stations assisting the stricken area. Amateur stations may also be designated to police the emergency frequencies, warning noncomplying stations that may be operating there. Other special conditions and rules may be set forth that will apply during the communications emergency. All amateurs must abide by any special terms or conditions stated in the FCC declaration of communication emergency.

G2C11 During a disaster in the US, which of the following emission modes must be used to obtain assistance?
  A. Only SSB
  B. Only SSB and CW
  C. Any mode
  D. Only CW

C   When normal communications are not available and the immediate safety of human life or protection of property is involved, all of the normal rules for an amateur station may be broken in order to obtain assistance. This means that any method of communication, any mode, on any frequency, and at any power output, may be used to communicate and resolve the emergency.

G2D Amateur auxiliary to the FCC's Compliance and Information Bureau; antenna orientation to minimize interference; HF operations, including logging practices

G2D01 What is the Amateur Auxiliary to the FCC's Compliance and Information Bureau?
    A. Amateur volunteers who are formally enlisted to monitor the airwaves for rules violations
    B. Amateur volunteers who conduct amateur licensing examinations
    C. Amateur volunteers who conduct frequency coordination for amateur VHF repeaters
    D. Amateur volunteers who use their station equipment to help civil defense organizations in times of emergency

A The purpose of the Amateur Auxiliary is to help ensure amateur self-regulation and see that amateurs follow the FCC rules properly. The Amateur Auxiliary volunteers deal only with amateur-to-amateur interference and improper operation. (Amateur volunteers who conduct licensing examinations are called VEs, Volunteer Examiners; amateurs in charge of frequency coordination for repeaters are called Frequency Coordinators; amateurs who help civil defense organizations in times of emergency are RACES members.)

G2D02 What are the objectives of the Amateur Auxiliary to the FCC's Compliance and Information Bureau?
    A. To conduct efficient and orderly amateur licensing examinations
    B. To encourage amateur self-regulation and compliance with the rules
    C. To coordinate repeaters for efficient and orderly spectrum usage
    D. To provide emergency and public safety communications

B The purpose of the Amateur Auxiliary is to help ensure amateur self-regulation and see that amateurs follow the FCC rules properly. The Amateur Auxiliary deals only with amateur-to-amateur interference and improper operation. (Amateur volunteers who conduct licensing examinations are called VEs, Volunteer Examiners; amateurs in charge of frequency coordination for repeaters are called Frequency Coordinators; amateurs who help civil defense organizations in times of emergency are RACES members.)

G2D03 Why are direction-finding "Fox Hunts" important to the Amateur Auxiliary?
  A. Fox Hunts compell amateurs to upgrade their licenses
  B. Fox Hunts provide an opportunity to practice direction-finding skills
  C. Someone always receives an FCC Notice of Apparent Liability (NAL) when a Fox Hunt is concluded
  D. Fox Hunts allow amateurs to work together with Environmental Protection Agencies

B Amateur Radio "Fox Hunts" are friendly hidden-transmitter competitions. Participants practice their radio direction-finding skills to track down the hidden transmitter – the "Fox." These skills become useful in tracking down harmful interference sources. The Amateur Auxilliary can use the Fox Hunters to document interference cases and report them to the proper enforcement bureau. Fox hunts also make everyone aware that there is a plan in place to find and eliminate an interference source.

G2D04 Which of the following is NOT an example of how "Direction Finding" skills help the Amateur Auxiliary?
  A. A good direction-finding team can pinpoint where any interference is originating from
  B. Good direction-finding skills lead to thorough records being created and forwarded to the proper enforcement bureau.
  C. Direction finding allows amateurs to operate outside our band limits
  D. Direction-finding drills make everyone aware that a plan is in place to eliminate interference.

C Amateur Radio "Fox Hunts" are friendly hidden-transmitter competitions. Participants practice their radio direction-finding skills to track down the hidden transmitter – the "Fox." These skills become useful in tracking down harmful interference sources. The Amateur Auxilliary can use the Fox Hunters to document interference cases and report them to the proper enforcement bureau. Fox hunts also make everyone aware that there is a plan in place to find and eliminate an interference source. The Amateur Radio Fox Hunters may help find an interference source outside the amateur bands, but that does not allow them to transmit outside of those band limits.

G2D05 What is an azimuthal map?
A. A map projection centered on the North Pole
B. A map projection centered on a particular location, used to determine the shortest path between points on the surface of the earth
C. A map that shows the angle at which an amateur satellite crosses the equator
D. A map that shows the number of degrees longitude that an amateur satellite appears to move westward at the equator with each orbit

B An azimuthal map, or azimuthal-equidistant projection map, is also called a great circle map. When this type map is centered on your location, a straight line is equivalent to stretching a string between two points on a globe, and will give you the shortest distance between two points. This type of map is used for determining the direction for short-path communications.

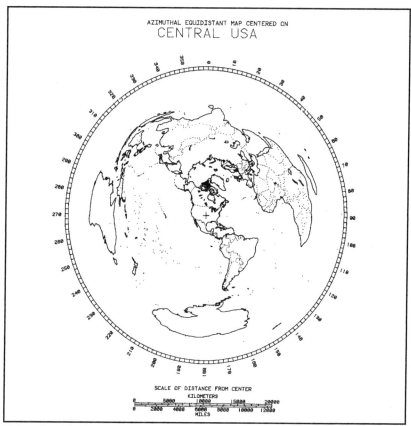

This azimuithal equidistant world map is centered on the central United States.

72   Subelement G2

G2D06 What is the most useful type of map to use when orienting a directional HF antenna toward a distant station?
A. Azimuthal
B. Mercator
C. Polar projection
D. Topographical

A An azimuthal map, or azimuthal-equidistant projection map, is also called a great circle map. When this type of map is centered on your location, a straight line is equivalent to stretching a string between two points on a globe, and will give you the shortest distance between two points. This type of map is used for orienting a directional HF antenna toward a distant station.

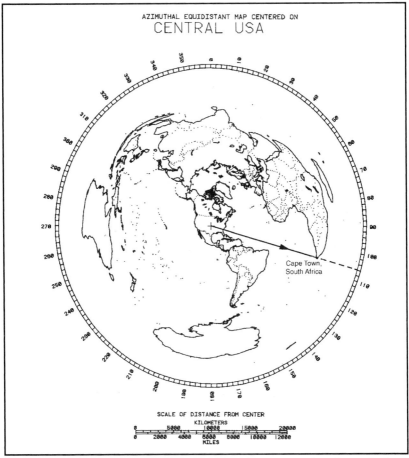

This azimuthal equidistant world map shows how an amateur in central United States would point a beam antenna for a contact with a station in Cape Town, South Africa.

Operating Procedures 73

G2D07 A directional antenna pointed in the long-path direction to another station is generally oriented how many degrees from its short-path heading?
- A. 45 degrees
- B. 90 degrees
- C. 180 degrees
- D. 270 degrees

C   The shortest direct route, or great-circle path between two points, is called the short-path. If a directional antenna is pointed in exactly the opposite direction, 180 degrees different from the short-path direction, communications can be attempted on the long-path. Long-path communication works best if the path is mostly over water.

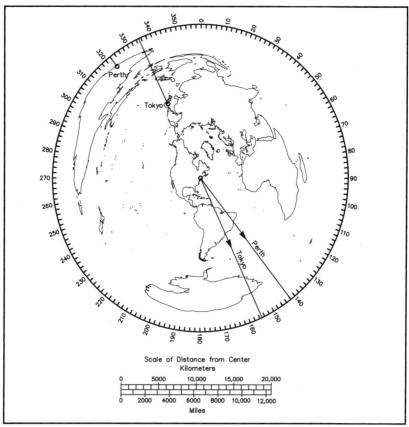

This azimuthal equidistant world map shows how W1AW in Newington, Connecticut would point a beam antenna to make a long-path contact with Tokyo, Japan or Perth, Australia. Notice that the paths in both cases lie almost entirely over water, rather than over land masses. This is the most likely condition for successful long-path communications.

G2D08 If a visiting amateur transmits from your station, which of these is NOT true?
  A. You must first give permission for the visiting amateur to use your station
  B. You must keep in your station log the call sign of the visiting amateur together with the time and date of transmissions
  C. The FCC may think that you were the station's control operator, unless your station records show otherwise
  D. You both are equally responsible for the proper operation of the station

B The FCC does not require you to keep a record (log) of your transmissions. It can be fun to keep a log, though, and then look back years later at the contacts you made. A log can also help document when your station was on the air and who was the control operator. You must give permission before a visiting amateur may operate your station. If you designate another amateur to be the control operator of your station, you both share the responsibility for the proper operation of the station. Unless your station records (log) show otherwise, the FCC will assume you were the control operator any time your station was operated.

G2D09 Why should I keep a log if the FCC doesn't require it?
  A. To help with your reply, if FCC requests information on who was control operator of your station for a given date and time
  B. Logs provide information (callsigns, dates & times of contacts) used for many operating contests and awards
  C. Logs are necessary to accurately verify contacts made weeks, months or years earlier, especially when completing QSL cards
  D. All of these choices

D The FCC does not require you to keep a record (log) of your transmissions. It can be fun to keep a log, though, and then look back years later at the contacts you made. Your log can provide information you need to enter contests or apply for operating awards. The information in your log will also help you fill out QSL cards for contacts made weeks, months or years earlier. A log can also help document when your station was on the air and who was the control operator, especially if the FCC requests that information.

Operating Procedures

G2D10 What is an advantage of keeping a paper log as well as a computer log?
  A. Paper logs will not accidentally erase like computer logs may
  B. There is no (computer start up) waiting time required to write information on paper
  C. All of these choices
  D. Paper logs can provide a back up for computer logs

C Many amateurs use computer logging programs to maintain records of their station operation. This can be convenient for contest operating and to quickly search for information about a past contact. There are also advantages to keeping a paper log, and many amateurs use the paper log as a backup for their computer log. Paper logs won't be accidentally erased like computer logs may be. With a paper log you won't have to wait for the computer to start up and run the logging program. With a pen and the paper log book in hand, you are ready to record data.

G2D11 What information is normally contained in a station log?
  A. Date and time of contact
  B. Band and/or frequency of the contact
  C. Call sign of station contacted and the RST signal report given
  D. All of these choices

D You can keep any information in your log that you would like to remember later. At a minimum, most amateurs keep a record of the date and time of each contact as well as the frequency or band of the contact. The station call sign and the RST signal report given and received are also normally recorded. Many amateurs also record the name of the other operator as well as his or her location.

G2E Third-party communications; ITU Regions; VOX operation

G2E01 What type of messages may be transmitted to an amateur station in a foreign country?
A. Messages of any type
B. Messages that are not religious, political, or patriotic in nature
C. Messages of a technical nature or personal remarks of relative unimportance
D. Messages of any type, but only if the foreign country has a third-party communications agreement with the US

C All communications with a foreign country (including those intended for a third-party) shall be limited to either technical (related to radio reception and/or transmission characteristics) or personal remarks that are not important enough to justify using public communication services. The goal is to avoid competition with the public telecommunication services of other countries. Remember that third-party communications are only allowed with countries that have third-party agreements with the US government. It doesn't make any difference whether you are transmitting in your own behalf or for a third party; none of the messages should be important enough that they really should go through commercial communication facilities.

G2E02 Which of the following statements is true of VOX operation?
A. The received signal is more natural sounding
B. Frequency spectrum is conserved
C. This mode allows "Hands Free" operation
D. The duty cycle of the transmitter is reduced

C The purpose of a voice operated transmit (VOX) circuit is to provide automatic transmit/receive (TR) switching within an amateur station. By simply speaking into the microphone, the antenna is connected to transmitter, the receiver is muted and the transmitter is activated. When you stop speaking, the VOX circuit switches everything back to receive. VOX allows hands-free operation.

G2E03 Which of the following user adjustable controls are usually associated with VOX circuitry?
  A. Anti-VOX
  B. VOX Delay
  C. VOX Sensitivity
  D. All of these choices are correct

D The purpose of a voice operated transmit (VOX) circuit is to provide automatic transmit/receive (TR) switching within an amateur station. The VOX Sensitivity control adjusts the level of microphone audio input needed to activate the circuit. VOX Delay adjusts how long the transmit circuit remains activated after you stop talking. The Anti-VOX control ensures that audio from the radio speaker won't activate the VOX circuit, putting the radio into transmit whenever a signal is received.

G2E04 What is the purpose of the VOX sensitivity control?
  A. To set the timing of transmitter activation
  B. To set the audio frequency range at which the transmitter activates
  C. To set the audio level at which the transmitter activates
  D. None of these choices is correct

C The purpose of a voice operated transmit (VOX) circuit is to provide automatic transmit/receive (TR) switching within an amateur station. The VOX Sensitivity control adjusts the level of microphone audio input needed to activate the circuit.

G2E05 What is the purpose of the Anti-VOX control?
  A. To prevent the received audio from activating the transmitter
  B. To prevent the transmitter from being activated by ambient or background noise
  C. To prevent activation of the transmitter during CW operation
  D. To override the function of other controls when the transmitter is used for frequency shift keying

A The purpose of a voice operated transmit (VOX) circuit is to provide automatic transmit/receive (TR) switching within an amateur station. The Anti-VOX control ensures that audio from the radio speaker won't activate the VOX circuit, putting the radio into transmit whenever a signal is received.

G2E06 In which International Telecommunication Union Region is the continental United States?
  A. Region 1
  B. Region 2
  C. Region 3
  D. Region 4

B   The International Telecommunication Union (ITU) Region 2 includes North and South America. ITU Region 1 is Africa, Europe, and Russia and surrounding countries. ITU Region 3 is Asia, Australia, and the South Pacific. There are only 3 ITU Regions.

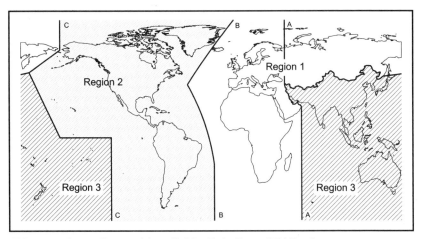

This map shows the world as divided into three ITU Regions.

G2E07 In which International Telecommunication Union Region are Europe and Africa?
  A. Region 1
  B. Region 2
  C. Region 3
  D. Region 4

A   The International Telecommunication Union (ITU) Region 1 is Africa, Europe, and Russia and surrounding countries. ITU Region 2 includes North and South America. ITU Region 3 is Asia, Australia, and the South Pacific. There are only 3 ITU Regions.

G2E08 In which International Telecommunication Union Region is Australia?

    A. Region 1
    B. Region 2
    C. Region 3
    D. Region 4

  C   The International Telecommunication Union (ITU) Region 3 is Asia, Australia, and the South Pacific. ITU Region 1 is Africa, Europe, and Russia and surrounding countries. ITU Region 2 includes North and South America. There are only 3 ITU Regions.

G2E09 Which of the following organizations are responsible for international regulation of the radio spectrum?

    A. The International Regulatory Commission
    B. The International Radio Union
    C. The International Telecommunication Union
    D. The International Frequency-Spectrum Commission

  C   The International Telecommunication Union (ITU) is an agency of the United Nations. The ITU is responsible for international regulation of the radio spectrum.

G2E10 What do the initials "ITU" stand for?

    A. Interstate Telecommunications Union
    B. International Telephony Union
    C. International Transmission Union
    D. International Telecommunication Union

  D   The International Telecommunication Union (ITU) is an agency of the United Nations. The ITU is responsible for international regulation of the radio spectrum.

G2E11 What is the circuit called that causes a transmitter to automatically transmit when an operator speaks into its microphone?

    A. VXO
    B. VOX
    C. VCO
    D. VFO

  B   The VOX (Voice-Operated Transmit) circuit has automatic switching between the transmitter and receiver. As soon as you begin speaking, the transmitter switches on, the antenna is connected to the transmitter (electronically), and the receiver is muted. When you stop speaking, the transmitter is turned off and the antenna is connected to the receiver (electronically).

G2F CW operating procedures, including procedural signals, Q signals and common abbreviations; full break-in; RTTY operating procedures, including procedural signals and common abbreviations and operating procedures for other digital modes, such as HF packet, AMTOR, PacTOR, G-TOR, Clover and PSK31

G2F01 Which of the following describes full break-in telegraphy?
A. Breaking stations send the Morse code prosign BK
B. Automatic keyers are used to send Morse code instead of hand keys
C. An operator must activate a manual send/receive switch before and after every transmission
D. Incoming signals are received between transmitted key pulses

D Full break-in telegraphy allows you to listen between your transmitted Morse code dots and dashes and between words. The advantage is that if you are sending message traffic, the receiving station can send back to you (break in) and stop you for repeats of missed words.

G2F02 In what segment of the 80-meter band do most RTTY transmissions take place?
A. 3580 - 3620 kHz
B. 3500 - 3525 kHz
C. 3700 - 3750 kHz
D. 3775 - 3825 kHz

A The FCC's rules list where RTTY is allowed; the band plans tell you where it is usually found. The band plan calls for operation between 3580 and 3620 kHz for 80 meters.

G2F03 In what segment of the 20-meter band do most RTTY transmissions take place?
A. 14.000 - 14.050 MHz
B. 14.070 - 14.095 MHz
C. 14.150 - 14.225 MHz
D. 14.275 - 14.350 MHz

B The FCC's rules list where RTTY is allowed; the band plans tell you where it is usually found. The band plan calls for operation between 14.070 to 14.095 MHz for 20 meters.

G2F04 What is the Baudot code?
- A. A 7-bit code, with start, stop and parity bits
- B. A 7-bit code in which each character has four mark and three space bits
- C. A 5-bit code, with additional start and stop bits
- D. A 6-bit code, with additional start, stop and parity bits

C The Baudot code is a 5-bit digital code used in teleprinter applications. In addition to the 5 data bits there are start and stop bits that separate the characters. Answer (A) describes ASCII, the American National Standard Code for Information Interchange, which is a code that is also used in computer applications.

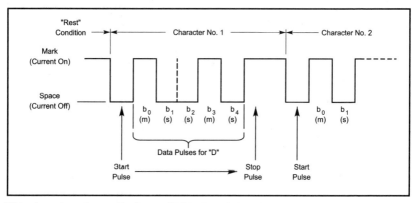

This drawing shows the letter "D" in Baudot code.

G2F05 What is the most common frequency shift for RTTY emissions in the amateur HF bands?
- A. 85 Hz
- B. 170 Hz
- C. 425 Hz
- D. 850 Hz

B RTTY operation on the HF bands uses frequency-shift keying (FSK) to convey the information. The transmitted frequency shifts between two frequencies, called the MARK and SPACE frequencies. The two frequencies used are normally 170 Hz apart.

G2F06 What is ASCII?

A. A 7-bit code, with additional start, stop and parity bits
B. A 7-bit code in which each character has four mark and three space bits
C. A 5-bit code, with additional start and stop bits
D. A 5-bit code in which each character has three mark and two space bits

A   ASCII stands for American National Standard Code for Information Interchange. This is a 7-bit digital code that is used in computer applications as well as radio teleprinter applications. In addition to the 7 data bits there are start and stop bits that separate the characters. There is also a parity bit that helps ensure that the character is received correctly. Because ASCII uses 7 information bits, more characters can be defined than with the 5-bit Baudot code.

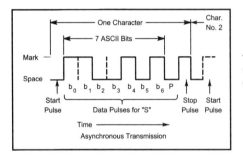

This drawing shows the letter "S" in ASCII, for asynchronous transmission.

G2F07 What are the two major AMTOR operating modes?

A. Mode AM and Mode TR
B. Mode A (ARQ) and Mode B (FEC)
C. Mode C (CRQ) and Mode D (DEC)
D. Mode SELCAL and Mode LISTEN

B   AMTOR stands for Amateur Teleprinting Over Radio. It combines a modified Baudot code with error-correcting capabilities. Mode A is called automatic repeat request (ARQ), and uses "handshaking" to communicate with a specific station. Mode B is called forward error correction (FEC), and is used to "broadcast" information for reception by more than one station. Mode B is also used to establish a contact, before a specific link-up is made. These are two different systems for ensuring that a communication is received correctly.

G2F08 Why are the string of letters R and Y (sent as "RYRYRYRY...") occasionally used at the beginning of RTTY transmissions?

    A. This allows time to 'tune in' a station prior to the actual message being sent
    B. To keep these commonly-used keys functional
    C. These are the important mark and space keys
    D. To make sure the transmitter is functional before sending a message

A   RTTY operators will occasionally send a string of the letters R and Y; RYRYRYRY. These two characters send a data stream that alternates between mark and space bits, and has a characteristic sound that can help other stations tune in the signal before the actual message text is sent.

G2F09 As one of several digital codes, what is the benefit of using AMTOR?

    A. Its error detection and correction properties
    B. Its data compression properties
    C. Its store and save properties
    D. It is a very narrow-bandwidth frequency efficient system

A   AMTOR — Amateur Teleprinting Over Radio — is a digital communications code that includes error detection and correction properties.

G2F10 What speed should you use when answering a CQ call using RTTY?

    A. Half the speed of the received signal
    B. The same speed as the received signal
    C. Twice the speed of the received signal
    D. Any speed, since RTTY systems adjust to any signal speed

B   Equipment settings as well as FCC Rules determine the rate at which digital signals such as RTTY can be sent. If a station sends RTTY at a certain rate, that is the speed you should use to reply because their equipment will be set up to receive at the same rate they used for transmitting.

G2F11 What does the abbreviation "RTTY" stand for?

    A. "Returning to you", meaning "your turn to transmit"
    B. Radioteletype
    C. A general call to all digital stations
    D. Morse code practice over the air

B   RTTY is a narrow-band, direct printing mode of telegraphy. RTTY stands for radioteletype.

Subelement G3

Radio Wave Propagation

Your General class exam will have 3 questions taken from the 3 groups of questions in this Radio-Wave Propagation subelement, G3A, G3B and G3C.

G3A Ionospheric disturbances; sunspots and solar radiation

G3A01 What can be done at an amateur station to continue communications during a sudden ionospheric disturbance?
- A. Try a higher frequency
- B. Try the other sideband
- C. Try a different antenna polarization
- D. Try a different frequency shift

A During a sudden ionospheric disturbance, you might try moving to a higher frequency to continue communication. Ultraviolet and X-ray radiation from the sun travels at the speed of light, reaching the Earth in about eight minutes. When this radiation reaches the Earth, the level of ionization in the atmosphere increases rapidly and D-layer absorption of radio waves increases significantly. Lower frequencies are affected more by absorption than higher frequencies.

G3A02 What effect does a sudden ionospheric disturbance have on the day-time ionospheric propagation of HF radio waves?
- A. It disrupts higher-latitude paths more than lower-latitude paths
- B. It disrupts signals on lower frequencies more than those on higher frequencies
- C. It disrupts communications via satellite more than direct communications
- D. None, only areas on the night side of the earth are affected

B Ultraviolet and X-ray radiation from the sun travels at the speed of light, reaching the Earth in about eight minutes. When this radiation reaches the Earth, the level of ionization in the atmosphere increases rapidly and D-layer absorption of radio waves increases significantly. Lower frequencies are affected more by absorption than higher frequencies. During a sudden ionospheric disturbance, you might try moving to a higher frequency.

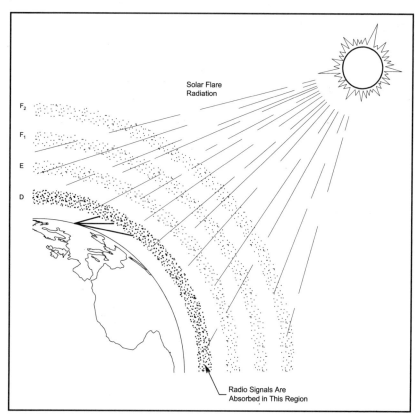

Approximately eight minutes after a solar flare occurs on the Sun, the ultraviolet and X-ray radiation released by the flare reaches the Earth. This radiation causes increased ionization and radio-wave absorption in the D region of the ionosphere.

G3A03 How long does it take the increased ultraviolet and X-ray radiation from solar flares to affect radio-wave propagation on the earth?
    A. The effect is almost instantaneous
    B. 1.5 minutes
    C. 8 minutes
    D. 20 to 40 hours

    C   Ultraviolet and X-ray radiation from the sun travels at the speed of light, reaching the Earth in about eight minutes. When this radiation reaches the Earth, the level of ionization in the atmosphere increases rapidly and D-layer absorption of radio waves increases significantly. Lower frequencies are affected more by absorption than higher frequencies. During a sudden ionospheric disturbance, you might try moving to a higher frequency.

G3A04 What is solar flux?
- A. The density of the sun's magnetic field
- B. The radio energy emitted by the sun
- C. The number of sunspots on the side of the sun facing the earth
- D. A measure of the tilt of the earth's ionosphere on the side toward the sun

B   Solar flux is the radio energy coming from the sun. High levels of solar energy produce greater ionization in the ionosphere. The advantage of using the solar-flux index is that the solar-flux measurement may be taken under any weather conditions (you do not have to have the sun visible, as you do to determine the sunspot number). The solar-flux measurement is taken daily (on 2800 MHz).

G3A05 What is the solar-flux index?
- A. A measure of solar activity that is taken annually
- B. A measure of solar activity that compares daily readings with results from the last six months
- C. Another name for the American sunspot number
- D. A measure of solar activity that is taken at a specific frequency

D   Solar flux is the radio energy coming from the sun. High levels of solar energy produce greater ionization in the ionosphere. The advantage of using the solar-flux index is that the solar-flux measurement may be taken under any weather conditions (you do not have to have the sun visible, as you do to determine the sunspot number). The solar-flux measurement is taken daily (on 2800 MHz).

G3A06 What is a geomagnetic disturbance?
- A. A sudden drop in the solar-flux index
- B. A shifting of the earth's magnetic pole
- C. Ripples in the ionosphere
- D. A dramatic change in the earth's magnetic field over a short period of time

D   Geomagnetic disturbances result when charged particles from a solar flare reach the Earth. When these charged particles reach the Earth's magnetic field, they are deflected towards the North and South poles. Radio communications along higher-latitude paths (latitudes greater than about 45 degrees) will be more affected than those close to the equator. These charged particles may make the F region seem to disappear or seem to split into many layers, degrading or completely blacking out long-distance radio communications.

G3A07 At which latitudes are propagation paths more sensitive to geomagnetic disturbances?
- A. Those greater than 45 degrees latitude
- B. Those between 5 and 45 degrees latitude
- C. Those near the equator
- D. All paths are affected equally

A Geomagnetic disturbances result when charged particles from a solar flare reach the Earth. When these charged particles reach the Earth's magnetic field, they are deflected towards the North and South poles. Radio communications along higher-latitude paths (latitudes greater than about 45 degrees) will be more affected than those close to the equator. These charged particles may make the F region seem to disappear or seem to split into many layers, degrading or completely blacking out long-distance radio communications.

G3A08 What can be the effect of a major geomagnetic storm on radio-wave propagation?
- A. Improved high-latitude HF propagation
- B. Degraded high-latitude HF propagation
- C. Improved ground-wave propagation
- D. Improved chances of UHF ducting

B Geomagnetic disturbances result when charged particles from a solar flare reach the Earth. When these charged particles reach the Earth's magnetic field, they are deflected towards the North and South poles. Radio communications along higher-latitude paths (latitudes greater than about 45 degrees) will be more affected than those close to the equator. These charged particles may make the F region seem to disappear or seem to split into many layers, degrading or completely blacking out long-distance radio communications.

G3A09 What phenomenon has the most effect on radio communication beyond ground-wave or line-of-sight ranges?
- A. Solar activity
- B. Lunar tidal effects
- C. The F1 region of the ionosphere
- D. The F2 region of the ionosphere

A Solar radiation ionizing the outer atmosphere is what causes the ionosphere to form. The ionosphere allows the penetration of radio waves or absorbs or bends them depending on the wavelength of the radio signal and the amount of ionization present. There are a number of regions of the ionosphere that have different effects on radio waves.

G3A10 Which two types of radiation from the sun influence propagation?
- A. Subaudible-frequency and audio-frequency emissions
- B. Electromagnetic and particle emissions
- C. Polar-region and equatorial emissions
- D. Infrared and gamma-ray emissions

B Electromagnetic (ultraviolet and X-ray radiation) emissions travel at the speed of light and take approximately eight minutes to reach the Earth from the sun. Charged particles, which result in a geomagnetic disturbance, take about 20 to 40 hours to reach the Earth.

G3A11 When sunspot numbers are high, what is the affect on radio communications?
- A. High-frequency radio signals are absorbed
- B. Frequencies above 300 MHz become usable for long-distance communication
- C. Long-distance communication in the upper HF and lower VHF range is enhanced
- D. High-frequency radio signals become weak and distorted

C When sunspot numbers are high, this means there is a significant amount of solar activity and there will be more ionization of the ionosphere. The more the ionosphere is ionized, the higher the frequency of radio signals that may be used for long-distance communication. During the peak of a sunspot cycle, the 20-meter (14 MHz) band will be open around the world even through the night. During an unusually good sunspot cycle, even the 6-meter (50 MHz) band can become usable for long-distance communication.

G3B Maximum usable frequency; propagation "hops"

G3B01 If the maximum usable frequency (MUF) on the path from Minnesota to France is 24 MHz, which band should offer the best chance for a successful contact?
- A. 10 meters
- B. 15 meters
- C. 20 meters
- D. 40 meters

B The closest amateur band with a frequency just below the maximum usable frequency (MUF) should offer the best chance for a successful contact. To determine the band, divide 300 by the frequency in MHz:

$$\text{Wavelength} = \frac{300}{\text{frequency (in MHz)}} = \frac{300}{24 \text{ MHz}} = 12.5 \text{ meters}$$

To find the closest amateur band with a frequency lower than this, select the next highest wavelength, which is the 15-meter band (21 MHz).

G3B02 If the maximum usable frequency (MUF) on the path from Ohio to Germany is 17 MHz, which band should offer the best chance for a successful contact?
- A. 80 meters
- B. 40 meters
- C. 20 meters
- D. 2 meters

C The closest amateur band with a frequency just below the maximum usable frequency (MUF) should offer the best chance for a successful contact. To determine the band, divide 300 by the frequency in MHz:

$$\text{Wavelength} = \frac{300}{\text{frequency (in MHz)}} = \frac{300}{17 \text{ MHz}} = 17.6 \text{ meters}$$

To find the closest amateur band with a frequency lower than this, select the next highest wavelength, which is the 20-meter band (14 MHz).

G3B03 If the HF radio-wave propagation (skip) is generally good on the 24-MHz and 28-MHz bands for several days, when might you expect a similar condition to occur?

A. 7 days later
B. 14 days later
C. 28 days later
D. 90 days later

C   It takes approximately 28 days for the sun to rotate on its axis. Since active areas on the sun may persist for four or five months, you can expect good propagation conditions to recur approximately every 28 days for four or five months.

G3B04 What is one way to determine if the maximum usable frequency (MUF) is high enough to support 28-MHz propagation between your station and western Europe?

A. Listen for signals on the 10-meter beacon frequency
B. Listen for signals on the 20-meter beacon frequency
C. Listen for signals on the 39-meter broadcast frequency
D. Listen for WWVH time signals on 20 MHz

A   28 MHz is the 10-meter band. To determine the band, divide 300 by the frequency in MHz:

$$\text{Wavelength} = \frac{300}{\text{frequency (in MHz)}} = \frac{300}{28 \text{ MHz}} = 10.7 \text{ meters}$$

Although this calculation does not come out exactly as 10 meters, it will help you identify the frequency as being part of the 10-meter band.

Beacons transmit signals so that amateur operators may evaluate propagation conditions. If you listen for a beacon station from Western Europe, you will be able to determine if the MUF is high enough for 10-meter communications to that area.

G3B05 What usually happens to radio waves with frequencies below the maximum usable frequency (MUF) when they are sent into the ionosphere?
- A. They are bent back to the earth
- B. They pass through the ionosphere
- C. They are completely absorbed by the ionosphere
- D. They are bent and trapped in the ionosphere to circle the Earth

A   The maximum usable frequency (MUF) relates to a particular desired destination. The MUF is the highest frequency that will allow the radio wave to reach its desired destination using E- or F-region propagation. There is no single MUF for a given transmitter location; it will vary depending on the direction and distance to the station you are attempting to contact. Frequencies lower than the MUF are generally bent back to to Earth. Frequencies higher than the MUF will pass through the ionosphere instead of being bent back to the Earth.

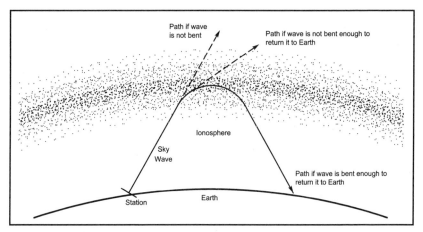

Radio waves are bent in the ionosphere, so they return to Earth far from the transmitter. If the radio wave is not bent (refracted) enough in the ionosphere, it will pass into space rather than returning to Earth.

G3B06 Where would you tune to hear beacons that would help you determine propagation conditions on the 20-meter band?
- A. 28.2 MHz
- B. 21.1 MHz
- C. 14.1 MHz
- D. 18.1 MHz

C   28.2 MHz is in the 10-meter band; 21.1 MHz is in the 15-meter band and 18.1 MHz is in the 17-meter band. For the 20-meter band, you want a frequency around 14 MHz. (To determine the frequency, divide 300 by the band in meters: 300 / 20 = 15.) The location of beacons within a band is determined by voluntary band plans.

G3B07 During periods of low solar activity, which frequencies are the least reliable for long-distance communication?
- A. Frequencies below 3.5 MHz
- B. Frequencies near 3.5 MHz
- C. Frequencies on or above 10 MHz
- D. Frequencies above 20 MHz

D The higher the frequency, the more ionization is needed in the ionosphere in order to refract (bend) the radio signal back to the Earth. When solar activity is low, the higher frequencies will pass out into space instead of being refracted back to Earth.

G3B08 At what point in the solar cycle does the 20-meter band usually support worldwide propagation during daylight hours?
- A. At the summer solstice
- B. Only at the maximum point of the solar cycle
- C. Only at the minimum point of the solar cycle
- D. At any point in the solar cycle

D Regardless of where you are in the 11-year solar cycle, you can depend on the 20-meter band for worldwide communication during the daylight hours. (During the height of the solar cycle, 20 meters will often stay open all night.)

G3B09 What is one characteristic of gray-line propagation?
- A. It is very efficient
- B. It improves local communications
- C. It is very poor
- D. It increases D-region absorption

A The gray-line (also called the twilight zone) is the band around the Earth that separates daylight from darkness. It is a fuzzy region, not a distinct line, because the Earth's atmosphere diffuses the light into the darkness. Propagation is very efficient along the gray-line because the D region, which absorbs HF signals, is disappearing rapidly on the sunset side and has not yet built up on the sunrise side.

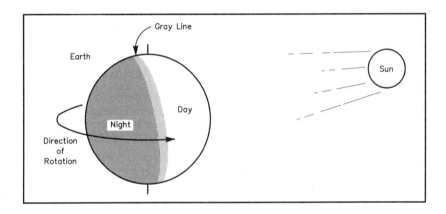

G3B10 What is the maximum distance along the Earth's surface that is normally covered in one hop using the F2 region?
- A. 180 miles
- B. 1200 miles
- C. 2500 miles
- D. None; the F2 region does not support radio-wave propagation

C The F region forms and decays in correlation with the passage of the sun. The F1 and F2 regions form when the F region splits into two parts under high ionization from the sun. They recombine into a single F region at night. The F2 region is the highest region of the ionosphere. It can reach as high as 300 miles at noon in the summertime. The more sunshine the F region receives, the more it will be ionized so it will be at its maximum ionization shortly after noon during the summertime. The ionization tapers off very gradually towards sunset and the F2 region remains usable throughout the night. Because the F2 region is the highest ionospheric region, it is the region mainly responsible for long-distance communications. A one-hop transmission can travel a maximum distance of about 2,500 miles using this F2 region.

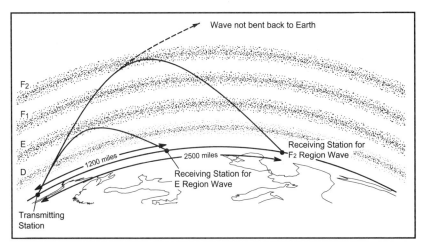

This drawing illustrates ionospheric propagation. Radio waves are refracted (bent) in the ionosphere and may return to Earth. If the radio waves are refracted to Earth from the F2 region, they may return to Earth about 2500 miles from the transmitting station. If the radio waves are refracted to Earth from the E region, they may return to Earth about 1200 miles from the transmitting station.

G3B11 What is the maximum distance along the Earth's surface that is normally covered in one hop using the E region?
- A. 180 miles
- B. 1200 miles
- C. 2500 miles
- D. None of these choices is correct

B   The E region of the ionosphere is the second lowest (above the D region). The E region appears at an altitude of about 70 miles above the Earth. The E region ionizes during the daytime, but the ionization does not last very long. Ionization of the E region is at a maximum around midday. During the daytime, a radio signal can travel a maximum distance of about 1,200 miles in one hop using the E region.

Radio Wave Propagation

G3C Height of ionospheric regions; critical angle and frequency; HF scatter

G3C01 What is the average height of maximum ionization of the E region?
- A. 45 miles
- B. 70 miles
- C. 200 miles
- D. 1200 miles

B The E region of the ionosphere is the second lowest (above the D region). The E region appears at an altitude of about 70 miles above the Earth. The E region ionizes during the daytime but the ionization does not last very long. Ionization of the E region is at a maximum around midday. During the daytime, a radio signal can travel a maximum distance of about 1200 miles in one hop using the E region.

G3C02 When can the F2 region be expected to reach its maximum height at your location?
- A. At noon during the summer
- B. At midnight during the summer
- C. At dusk in the spring and fall
- D. At noon during the winter

A The F region forms and decays in correlation with the passage of the sun. The F1 and F2 regions form when the F region splits into two parts under high ionization from the sun. They recombine into a single F region at night. The F2 region is the highest region of the ionosphere. It can reach as high as 300 miles at noon in the summertime. The more sunshine the F region receives the more it will be ionized, so it will be at its maximum ionization shortly after noon during the summertime. The ionization tapers off very gradually towards sunset. Because the F2 region is the highest ionospheric region, it is the region mainly responsible for long-distance communications. A one-hop transmission can travel a maximum distance of about 2,500 miles using this F2 region.

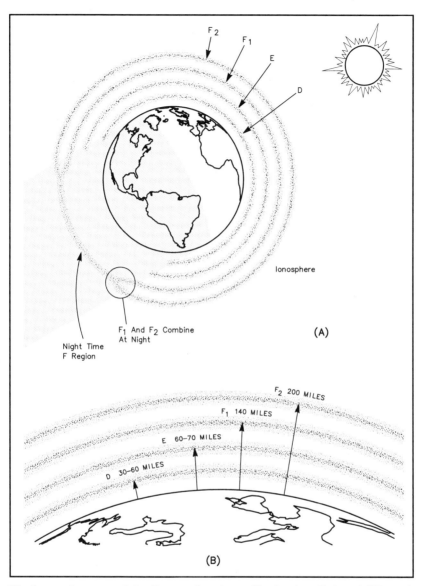

The ionosphere consists of several regions of ionized particles at different heights above the Earth. At night, the D and E regions disappear and the F1 and F2 regions combine to form a single F region.

G3C03 Why is the F2 region mainly responsible for the longest-distance radio-wave propagation?
- A. Because it exists only at night
- B. Because it is the lowest ionospheric region
- C. Because it is the highest ionospheric region
- D. Because it does not absorb radio waves as much as other ionospheric regions

    C   The F region forms and decays in correlation with the passage of the sun. The F1 and F2 regions form when the F region splits into two parts under high ionization from the sun. They recombine into a single F region at night. The F2 region is the highest region of the ionosphere. It can reach as high as 300 miles at noon in the summertime. Because the F2 region is the highest ionospheric region, it is the region mainly responsible for long-distance communications. A one-hop transmission can travel a maximum distance of about 2,500 miles using this F2 region. The more sunshine the F region receives the more it will be ionized, so it will be at its maximum ionization shortly after noon during the summertime. The ionization tapers off very gradually towards sunset.

G3C04 What is the "critical angle" as used in radio-wave propagation?
- A. The lowest takeoff angle that will return a radio wave to the earth under specific ionospheric conditions
- B. The compass direction of a distant station
- C. The compass direction opposite that of a distant station
- D. The highest takeoff angle that will return a radio wave to the earth under specific ionospheric conditions

    D   For each frequency there is a maximum angle at which the radio wave can leave the antenna and still be refracted back by the ionosphere instead of simply passing through it and proceeding out into space. The critical angle will change depending on the ionization of the ionosphere.

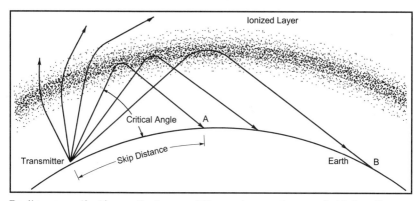

Radio waves that leave the transmitting antenna at an angle higher than the critical angle are not refracted enough to return to Earth. A radio wave at the critical angle will return to Earth. The lowest-angle wave will return to Earth farther away than the wave at the critical angle. This illustrates the importance of low radiation angles for working DX.

G3C05 What is the main reason the 160-, 80- and 40-meter amateur bands tend to be useful only for short-distance communications during daylight hours?
  A. Because of a lack of activity
  B. Because of auroral propagation
  C. Because of D-region absorption
  D. Because of magnetic flux

C Think of the D region as the *Darned Daylight* region. The D region does not bend high frequency signals back to Earth. What it does instead is absorb energy from radio waves. Lower frequencies (longer wavelengths such as 160, 80 and 40-meters) are absorbed more than higher frequencies. The ionization created by the sunlight does not last very long in the D region; it will disappear by sunset.

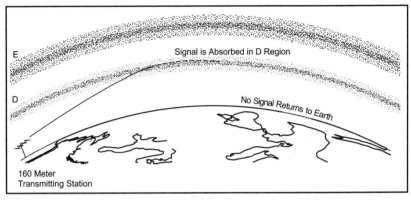

The D region of the ionosphere absorbes energy from radio waves. Lower-frequency radio waves don't make it all the way through the D region, so the waves do not return to Earth. Higher-frequency waves travel through the D region, and are then refracted (bent) back to Earth.

G3C06 What is a characteristic of HF scatter signals?
  A. High intelligibility
  B. A wavering sound
  C. Reversed modulation
  D. Reversed sidebands

B   The area between the farthest reach of ground-wave propagation and the point where signals are refracted back from the ionosphere (sky-wave propagation) is called the skip zone. Since some of the transmitted signal is scattered in the atmosphere, communication may be possible in the skip zone by the use of scatter signals. The amount of signal scattered in the atmosphere will be quite small and the signal received in the skip zone will arrive from several radio-wave paths. This tends to produce a distorted signal with a fluttering or wavering sound.

G3C07 What makes HF scatter signals often sound distorted?
A. Auroral activity and changes in the earth's magnetic field
B. Propagation through ground waves that absorb much of the signal
C. The state of the E-region at the point of refraction
D. Energy scattered into the skip zone through several radio-wave paths

D The area between the farthest reach of ground-wave propagation and the point where signals are refracted back from the ionosphere (sky-wave propagation) is called the skip zone. Since some of the transmitted signal is scattered in the atmosphere, communication may be possible in the skip zone by the use of scatter signals. The amount of signal scattered in the atmosphere will be quite small and the signal received in the skip zone will arrive from several radio-wave paths. This tends to produce a distorted signal with a fluttering or wavering sound.

G3C08 Why are HF scatter signals usually weak?
A. Only a small part of the signal energy is scattered into the skip zone
B. Auroral activity absorbs most of the signal energy
C. Propagation through ground waves absorbs most of the signal energy
D. The F region of the ionosphere absorbs most of the signal energy

A The area between the farthest reach of ground-wave propagation and the point where signals are refracted back from the ionosphere (sky-wave propagation) is called the skip zone. Since some of the transmitted signal is scattered in the atmosphere, communication may be possible in the skip zone by the use of scatter signals. The amount of signal scattered in the atmosphere will be quite small and the signal received in the skip zone will arrive from several radio-wave paths. This tends to produce a distorted signal with a fluttering or wavering sound.

G3C09 What type of radio-wave propagation allows a signal to be detected at a distance too far for ground-wave propagation but too near for normal sky-wave propagation?
A. Ground wave
B. Scatter
C. Sporadic-E skip
D. Short-path skip

B   The area between the farthest reach of ground-wave propagation and the point where signals are refracted back from the ionosphere (sky-wave propagation) is called the skip zone. Since some of the transmitted signal is scattered in the atmosphere, communication may be possible in the skip zone by the use of scatter signals. The amount of signal scattered in the atmosphere will be quite small and the signal received in the skip zone will arrive from several radio-wave paths. This tends to produce a distorted signal with a fluttering or wavering sound.

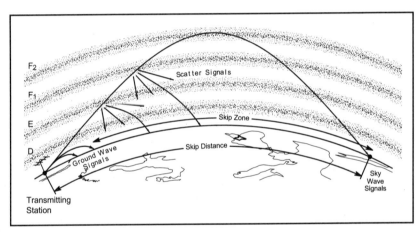

The skip zone is an area between the farthest reaches of ground-wave propagation and the closest return of sky waves from the ionosphere. Signals scattered in the ionosphere can help "fill in" the skip zone with weak, sometimes distorted signals.

G3C10 When does scatter propagation on the HF bands most often occur?

A. When the sunspot cycle is at a minimum and D-region absorption is high
B. At night
C. When the F1 and F2 regions are combined
D. When communicating on frequencies above the maximum usable frequency (MUF)

D The area between the farthest reach of ground-wave propagation and the point where signals are refracted back from the ionosphere (sky-wave propagation) is called the skip zone. Since some of the transmitted signal is scattered in the atmosphere, communication may be possible in the skip zone by the use of scatter signals. The amount of signal scattered in the atmosphere will be quite small and the signal received in the skip zone will arrive from several radio-wave paths. This tends to produce a distorted signal with a fluttering or wavering sound. Frequencies above the maximum usable frequency (MUF) normally pass through the ionosphere out into space rather than being bent back, although atmospheric scatter from the ionosphere will sometimes allow communication on these frequencies.

G3C11 What is one way the vertical incidence critical frequency measurement might be used?

A. It can be used to measure noise arriving vertically from outer space
B. It can be used to measure the vertical angle at which to set your beam antenna
C. It can be used to determine the size of coronal holes in the ionosphere
D. It can be used to determine the maximum usable frequency for long-distance communication at the time of measurement

D The vertical incidence critical frequency measurement helps determine the maximum usable frequency (MUF) for a certain propagation path. Amateurs usually determine the MUF by using computer software or published charts.

Subelement G4

Amateur Radio Practices

There will be 5 questions on your General class exam from the Amateur Radio Practices subelement. Those questions will be taken from the 5 groups labeled G4A through G4E, printed in this chapter.

G4A Two-tone test; electronic TR switch; amplifier neutralization

G4A01 What kind of input signal is used to test the amplitude linearity of a single-sideband phone transmitter while viewing the output on an oscilloscope?
  A. Normal speech
  B. An audio-frequency sine wave
  C. Two audio-frequency sine waves
  D. An audio-frequency square wave

C It is common to test the amplitude linearity of a single-sideband transmitter by injecting two audio tones of equal level into the microphone jack, then observing the pattern made on an oscilloscope. In order to get meaningful results, the two tones must not be harmonically related to each other (like 1,000 and 2,000 kHz). Of course, in order for the audio frequencies to display on the oscilloscope, they must be within the audio passband of the transmitter.

G4A02 When testing the amplitude linearity of a single-sideband transmitter, what kind of audio tones are fed into the microphone input and on what kind of instrument is the transmitter output observed?
    A. Two harmonically related tones are fed in, and the output is observed on an oscilloscope
    B. Two harmonically related tones are fed in, and the output is observed on a distortion analyzer
    C. Two non harmonically related tones are fed in, and the output is observed on an oscilloscope
    D. Two non harmonically related tones are fed in, and the output is observed on a distortion analyzer

    C   It is common to test the amplitude linearity of a single-sideband transmitter by injecting two audio tones of equal level into the microphone jack, then observing the pattern made on an oscilloscope. In order to get meaningful results, the two tones must not be harmonically related to each other (like 1,000 and 2,000 kHz). Of course, in order for the audio frequencies to display on the oscilloscope, they must be within the audio passband of the transmitter.

G4A03 What audio frequencies are used in a two-tone test of the linearity of a single-sideband phone transmitter?
    A. 20 Hz and 20-kHz tones must be used
    B. 1200 Hz and 2400 Hz tones must be used
    C. Any two audio tones may be used, but they must be within the transmitter audio passband, and must be harmonically related
    D. Any two audio tones may be used, but they must be within the transmitter audio passband, and should not be harmonically related

    D   It is common to test the amplitude linearity of a single-sideband transmitter by injecting two audio tones of equal level into the microphone jack, then observing the pattern made on an oscilloscope. In order to get meaningful results, the two tones must not be harmonically related to each other (like 1,000 and 2,000 kHz). Of course, in order for the audio frequencies to display on the oscilloscope, they must be within the audio passband of the transmitter.

G4A04 What measurement can be made of a single-sideband phone transmitter's amplifier by performing a two-tone test using an oscilloscope?
  A. Its percent of frequency modulation
  B. Its percent of carrier phase shift
  C. Its frequency deviation
  D. Its linearity

D It is common to test the amplitude linearity of a single-sideband transmitter by injecting two audio tones of equal level into the microphone jack, then observing the pattern made on an oscilloscope.

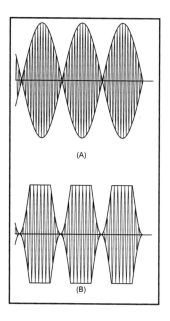

The two-tone test waveform at A is from a properly operating SSB transmitter. Note the smooth sine-wave shape of the curves, especially as they cross the zero axis. The two-tone test waveform at B is from a transmitter that has severe flattopping and crossover distortion. Notice the different shape of the waveform at the zero crossing, as compared to the waveform shown at A.

G4A05 At what point in an HF transceiver block diagram would an electronic TR switch normally appear?
  A. Between the transmitter and low-pass filter
  B. Between the low-pass filter and antenna
  C. At the antenna feed point
  D. At the power supply feed point

A   The TR switch is located between the transmitter/receiver and the low-pass filter, which is placed nearest the antenna to minimize television interference.

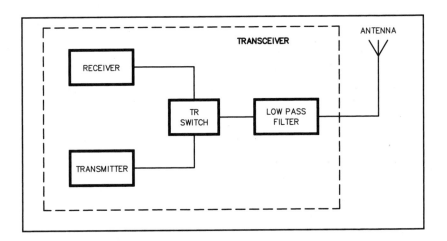

G4A06 Why is an electronic TR switch preferable to a mechanical one?
  A. It allows greater receiver sensitivity
  B. Its circuitry is simpler
  C. It has a higher operating speed
  D. It allows cleaner output signals

C   An electronic TR switch can operate so fast that the operator is able to hear other people's signals in between the dots and dashes of his or her own Morse code transmission.

G4A07 As a power amplifier is tuned, what reading on its grid-current meter indicates the best neutralization?
- A. A minimum change in grid current as the output circuit is changed
- B. A maximum change in grid current as the output circuit is changed
- C. Minimum grid current
- D. Maximum grid current

A Neutralization of a power amplifier is a technique designed to minimize or cancel the effects of positive feedback. Positive feedback occurs when the output signal gets fed back into the input in phase with the input signal. As a result, the amplifier is said to oscillate. You can demonstrate this principle by observing the squeal that occurs when you hold a microphone in front of a speaker. Neutralization consists of feeding a portion of the amplifier output back to the input, 180 degrees out of phase with the input, to cancel the positive feedback. One way of determining the proper amount of out-of-phase input is to observe the change in the grid current as the output circuit tuning is changed. There should be a minimum change in grid current as the output circuit is changed.

G4A08 Why is neutralization necessary for some vacuum-tube amplifiers?
- A. To reduce the limits of loaded Q
- B. To reduce grid-to-cathode leakage
- C. To cancel AC hum from the filament transformer
- D. To cancel oscillation caused by the effects of interelectrode capacitance

D Neutralization of a power amplifier is a technique designed to minimize or cancel the effects of positive feedback. Positive feedback occurs when the output signal gets fed back into the input in phase with the input signal. As a result, the amplifier is said to oscillate. You can demonstrate this principle by observing the squeal that occurs when you hold a microphone in front of a speaker. Even a small amount of capacitance between the electrodes inside the tube can cause the circuit to oscillate. Neutralization consists of feeding a portion of the amplifier output back to the input, 180 degrees out of phase with the input, to cancel the feedback. One way of determining the proper amount of out-of-phase input is to observe the change in the grid current as the output circuit is changed. There should be a minimum change in grid current as the output circuit is changed.

G4A09 In a properly neutralized RF amplifier, what type of feedback is used?
A. 5%
B. 10%
C. Negative
D. Positive

C Neutralization of a power amplifier is a technique designed to minimize or cancel the effects of positive feedback. Positive feedback occurs when the output signal gets fed back into the input in phase with the input signal. As a result, the amplifier is said to oscillate. You can demonstrate this principle by observing the squeal that occurs when you hold a microphone in front of a speaker. Neutralization consists of feeding a portion of the amplifier output back to the input, 180 degrees out of phase with the input, to cancel the feedback. This is called negative feedback.

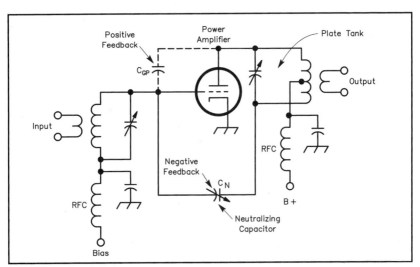

In most vacuum-tube amplifiers, some form of neutralization must be used to cancel positive feedback from the plate to the grid. In this circuit, a neutralizing capacitor (C_N) is added to counteract the effects of the grid-to-plate capacitance (shown as C_{GP}). The signal at the bottom of the tank circuit is 180° out of phase with the signal at the top of the tank, and careful adjustment of C_N will effectively cancel the positive feedback.

G4A10 What does a neutralizing circuit do in an RF amplifier?
  A. It controls differential gain
  B. It cancels the effects of positive feedback
  C. It eliminates AC hum from the power supply
  D. It reduces incidental grid modulation

B Neutralization of a power amplifier is a technique designed to minimize or cancel the effects of positive feedback. Positive feedback occurs when the output signal gets fed back into the input in phase with the input signal. As a result, the amplifier is said to oscillate. You can demonstrate this principle by observing the squeal that occurs when you hold a microphone in front of a speaker. Neutralization consists of feeding a portion of the amplifier output back to the input, 180 degrees out of phase with the input, to cancel the feedback.

G4A11 What is the reason for neutralizing the final amplifier stage of a transmitter?
  A. To limit the modulation index
  B. To eliminate self oscillations
  C. To cut off the final amplifier during standby periods
  D. To keep the carrier on frequency

B Neutralization of a power amplifier is a technique designed to minimize or cancel the effects of positive feedback. Positive feedback occurs when the output signal gets fed back into the input in phase with the input signal. As a result, the amplifier is said to oscillate. You can demonstrate this principle by observing the squeal that occurs when you hold a microphone in front of a speaker. Neutralization consists of feeding a portion of the amplifier output back to the input, 180 degrees out of phase with the input, to cancel the feedback. Neutralization prevents the amplifier from going into self oscillation.

G4B Test equipment: oscilloscope; signal tracer; antenna noise bridge; monitoring oscilloscope; field-strength meters

G4B01 What item of test equipment contains horizontal- and vertical-channel amplifiers?
  A. An ohmmeter
  B. A signal generator
  C. An ammeter
  D. An oscilloscope

D An oscilloscope has two channels. This allows two signals to be compared to each other by applying one signal to the vertical input and the other to the horizontal input, resulting in a characteristic pattern on the screen.

G4B02 What is a digital oscilloscope?
 A. An oscilloscope used only for signal tracing in digital circuits
 B. An oscilloscope used only for troubleshooting computers
 C. An oscilloscope used only for troubleshooting switching power supply circuits
 D. An oscilloscope designed around digital technology rather than analog technology

D Modern electronics is making more use of digital circuits. They represent the measured physical quantities with a series of numbers, or digits. Using inexpensive digital integrated circuits, a digital oscilloscope can store, change and compare waveforms.

G4B03 How would a signal tracer normally be used?
 A. To identify the source of radio transmissions
 B. To make exact drawings of signal waveforms
 C. To show standing wave patterns on open-wire feed-lines
 D. To identify an inoperative stage in a receiver

D A signal tracer is any device that reacts to RF or AF energy, and is used to trace the path of a signal through a circuit. A signal tracer can help identify an inoperative stage in a receiver.

G4B04 Why would you use a noise bridge?
 A. To measure the noise figure of an antenna or other electrical circuit
 B. To measure the impedance of an antenna or other electrical circuit
 C. To cancel electrical noise picked up by an antenna
 D. To tune out noise in a receiver

B A noise bridge is a device that applies a wideband noise signal to a circuit to test the impedance of the circuit. A receiver on the desired frequency is also placed in the circuit. Adjust the noise bridge to minimize the noise received, and then observe the readings on the noise bridge controls. The markings on the noise bridge controls indicate the impedance of the circuit.

G4B05 How is a noise bridge normally used?
- A. It is connected at an antenna's feed point and reads the antenna's noise figure
- B. It is connected between a transmitter and an antenna and is tuned for minimum SWR
- C. It is connected between a receiver and an antenna of unknown impedance and is tuned for minimum noise
- D. It is connected between an antenna and ground and is tuned for minimum SWR

C A noise bridge is a device that applies a wideband noise signal to a circuit to test the impedance of the circuit. A receiver on the desired frequency is also placed in the circuit. Adjust the noise bridge to minimize the noise received, and then observe the readings on the noise bridge controls. The markings on the noise bridge controls indicate the impedance of the circuit.

G4B06 What is the best instrument to use to check the signal quality of a CW or single-sideband phone transmitter?
- A. A monitoring oscilloscope
- B. A field-strength meter
- C. A sidetone monitor
- D. A signal tracer and an audio amplifier

A An oscilloscope visually displays a signal waveform. This allows you to observe the shape of the CW signal (referred to as the CW envelope) noting, for example, the rise and fall rate of the signal. It also allows you to observe problems like flattopping (caused by overmodulation) on your SSB signal.

G4B07 What signal source is connected to the vertical input of a monitoring oscilloscope when checking the quality of a transmitted signal?
- A. The IF output of a monitoring receiver
- B. The audio input of the transmitter
- C. The RF signals of a nearby receiving antenna
- D. The RF output of the transmitter

D When the RF output of a transmitter is connected to the vertical channel of an oscilloscope, the oscilloscope visually displays a signal waveform. This allows you to check for signal distortion such as flattopping (caused by overmodulation).

G4B08 What instrument can be used to determine the horizontal radiation pattern of an antenna?
- A. A field-strength meter
- B. A grid-dip meter
- C. An oscilloscope
- D. A signal tracer and an audio amplifier

A A field-strength meter makes a relative measurement of the intensity of the field being radiated from an antenna. By placing the field-strength meter in different locations around the antenna, you can determine the directivity of the antenna. This will give a relative indication of the horizontal radiation pattern of the antenna.

G4B09 How is a field-strength meter normally used?
- A. To determine the standing-wave ratio on a transmission line
- B. To check the output modulation of a transmitter
- C. To monitor relative RF output
- D. To increase average transmitter output

C A field-strength meter makes a relative measurement of the intensity of the field being radiated from an antenna. Some amateurs keep a field-strength meter in their shack to monitor the relative RF output from their station.

G4B10 What simple instrument may be used to monitor relative RF output during antenna and transmitter adjustments?
- A. A field-strength meter
- B. An antenna noise bridge
- C. A multimeter
- D. A metronome

A A field-strength meter makes a relative measurement of the intensity of the field being radiated from an antenna. Some amateurs keep a field-strength meter in their shack to monitor the relative RF output from their station. This can be especially handy when you are making antenna or transmitter adjustments.

G4B11 By how many times must the power output of a transmitter be increased to raise the S-meter reading on a nearby receiver from S8 to S9?

    A. Approximately 2 times
    B. Approximately 3 times
    C. Approximately 4 times
    D. Approximately 5 times

C   A theoretical, ideal S-meter operates on a logarithmic scale, providing a one-unit indicated change for a four-times increase or decrease in power. (This is a 6-dB change in power.) In real life, S-meters are only calibrated to this standard in the middle of their range (if at all).

G4C Audio rectification in consumer electronics; RF ground

G4C01 What devices would you install in home-entertainment systems to reduce or eliminate audio-frequency interference?

    A. Bypass inductors
    B. Bypass capacitors
    C. Metal-oxide varistors
    D. Bypass resistors

B   If radio frequency interference is entering a home audio system through external speaker wires, bypassing the wires to the chassis through capacitors at the speaker terminals is usually effective. (In practice, it's a little more complicated than this, though. With transistor audio amplifiers you may need to use RF chokes in series with the speaker leads, as well as bypass capacitors. See the *ARRL Handbook* and *The ARRL RFI Book* for more information on finding and fixing RFI problems.)

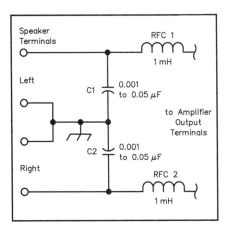

This diagram shows a filter that can be used at the output of some transistor audio amplifiers. The capacitors bypass the undesired signals to ground.
C1, C2 — 0.001 to 0.05-µF disc ceramic, 1.4-kV capacitors.
RFC1, RFC2 — 24 turns of number 18 AWG enameled wire, close spaced and wound on a ¼-inch diameter form (such as a pencil).

Amateur Radio Practices   115

G4C02 What should be done if a properly operating amateur station is the cause of interference to a nearby telephone?
- A. Make internal adjustments to the telephone equipment
- B. Install RFI filters at the affected telephone
- C. Stop transmitting whenever the telephone is in use
- D. Ground and shield the local telephone distribution amplifier

B If you or your neighbor are experiencing telephone interference, you first need to determine who owns the phone and who maintains the phone lines inside the home. If the phones are rented from the phone company and they maintain the lines inside the home contact the local telephone company. They have inductors and other filter components to help eliminate the problem. If the customer owns the phones, have a qualified repair technician do the work of installing the components, in case there is a problem later.

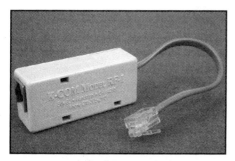

This telephone interference filter, manufactured by K-Com, has modular connectors to simplifiy installation right at the telephone.

G4C03 What sound is heard from a public-address system if audio rectification of a nearby single-sideband phone transmission occurs?
- A. A steady hum whenever the transmitter's carrier is on the air
- B. On-and-off humming or clicking
- C. Distorted speech from the transmitter's signals
- D. Clearly audible speech from the transmitter's signals

C A public-address system can sometimes rectify (detect) RF signals in much the same way that a radio does. The audio signal is then amplified by the audio device resulting in interference. The amateur's voice will be heard but it will be highly distorted.

G4C04 What sound is heard from a public-address system if audio rectification of a nearby CW transmission occurs?
- A. On-and-off humming or clicking
- B. Audible, possibly distorted speech
- C. Muffled, severely distorted speech
- D. A steady whistling

A A public-address system can sometimes rectify (detect) RF signals in much the same way that a radio does. The audio signal is then amplified by the audio device resulting in interference. An amateur's CW transmission will be heard as on-and-off humming or clicking.

G4C05 How can you minimize the possibility of audio rectification of your transmitter's signals?
- A. By using a solid-state transmitter
- B. By using CW emission only
- C. By ensuring that all station equipment is properly grounded
- D. By installing bypass capacitors on all power supply rectifiers

C A public-address system can sometimes rectify (detect) RF signals in much the same way that a radio does. The audio signal is then amplified by the audio device resulting in interference. You can minimize the possibility of sending unwanted signals by ensuring that all station equipment is properly grounded.

G4C06 If your third-floor amateur station has a ground wire running 33 feet down to a ground rod, why might you get an RF burn if you touch the front panel of your HF transceiver?
- A. Because the ground rod is not making good contact with moist earth
- B. Because the transceiver's heat-sensing circuit is not working to start the cooling fan
- C. Because of a bad antenna connection, allowing the RF energy to take an easier path out of the transceiver through you
- D. Because the ground wire is a resonant length on several HF bands and acts more like an antenna than an RF ground connection

D The purpose of a ground wire is to make sure that all unwanted signals are shorted to ground instead of being radiated. If the ground wire is long enough to be resonant on one or more bands, however, it can bring some of the radio signal energy into the station and put RF voltages on your equipment chassis.

G4C07 Which of the following is NOT an important reason to have a good station ground?
- A. To reduce the cost of operating a station
- B. To reduce electrical noise
- C. To reduce interference
- D. To reduce the possibility of electric shock

A The purpose of a ground wire is to short any undesired RF radiation, electrical noise, or even electrical power, to ground, and to prevent electrical shock. A ground wire will NOT reduce the cost of operating your station.

G4C08 What is one good way to avoid stray RF energy in your amateur station?
- A. Keep the station's ground wire as short as possible
- B. Use a beryllium ground wire for best conductivity
- C. Drive the ground rod at least 14 feet into the ground
- D. Make a couple of loops in the ground wire where it connects to your station

A The purpose of a ground wire is to make sure that all unwanted signals are shorted to ground instead of being radiated. If the ground wire is long, however, it can actually serve as an antenna, inadvertently radiating the undesired signals or bringing stray RF energy into your station.

G4C09 Which of the following statements about station grounding is NOT true?
- A. Braid from RG-213 coaxial cable makes a good conductor to tie station equipment together into a station ground
- B. Only transceivers and power amplifiers need to be tied into a station ground
- C. According to the National Electrical Code, there should be only one grounding system in a building
- D. The minimum length for a good ground rod is 8 feet

B The purpose of a ground wire is to short any undesired RF radiation, electrical noise, or even electrical power, to ground. It is important to include each piece of equipment in the grounding network, not just transceivers and amplifiers. The National Electrical Code lists specific grounding requirements, including the need to have only one grounding system for a building. (One way to satisfy this requirement is to connect your RF station ground to the electrical ground connection — not just the grounded wire in your station wiring.) The minimum length for a ground rod is 8 feet.

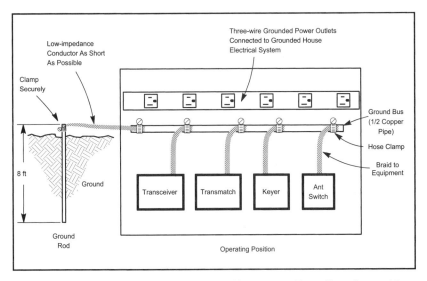

You can make an effective station ground by connecting all equipment to a ground bus. A length of ½ inch copper pipe along the back of your operating desk or table makes a good ground bus. Heavy copper braid, such as the outer braid from RG-8 coaxial cable makes a good, flexible strap to connect each piece of equipment to the ground bus. The whole system then connects to a good earth ground — an 8-foot ground rod located as close to the station as possible is the minimum ground you should use. More than one ground rod may be needed in some locations.

G4C10 Which of the following statements about station grounding is true?

    A. The chassis of each piece of station equipment should be tied together with high-impedance conductors
    B. If the chassis of all station equipment is connected with a good conductor, there is no need to tie them to an earth ground
    C. RF hot spots can occur in a station located above the ground floor if the equipment is grounded by a long ground wire
    D. A ground loop is an effective way to ground station equipment

    C   The purpose of a ground wire is to make sure that all unwanted signals are shorted to ground instead of being radiated. If the ground wire is long enough to be resonant on one or more bands, however, it can bring some of the radio signal energy into the station and put RF voltages on your equipment chassis. This results in "RF hot spots," a common problem with stations located above the ground floor.

Amateur Radio Practices   119

G4C11 Which of the following is NOT covered in the National Electrical Code?
- A. Minimum conductor sizes for different lengths of amateur antennas
- B. The size and composition of grounding conductors
- C. Electrical safety inside the ham shack
- D. The RF exposure limits of the human body

D The National Electrical Code covers the wiring of electrical devices, including minimum conductor sizes for antennas, grounding conductors and general electrical safety. It does not deal with RF radiation exposure limits. The FCC Rules specify RF radiation exposure limits.

G4D Speech processors; PEP calculations; wire sizes and fuses

G4D01 What is the reason for using a properly adjusted speech processor with a single-sideband phone transmitter?
- A. It reduces average transmitter power requirements
- B. It reduces unwanted noise pickup from the microphone
- C. It improves voice-frequency fidelity
- D. It improves signal intelligibility at the receiver

D A speech processor can improve signal intelligibility by raising average power without increasing peak envelope power (PEP). It does this by bringing up low signal levels while not increasing high signal levels. As a result, the average signal level is higher. A speech processor does not increase the output PEP.

G4D02 If a single-sideband phone transmitter is 100% modulated, what will a speech processor do to the transmitter's power?
- A. It will increase the output PEP
- B. It will add nothing to the output PEP
- C. It will decrease the peak power output
- D. It will decrease the average power output

B A speech processor can improve signal intelligibility by raising average power without increasing peak envelope power (PEP). It does this by bringing up low signal levels while not increasing high signal levels. As a result, the average signal level is higher. A speech processor does not increase the output PEP.

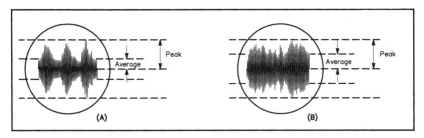

A typical SSB voice-modulated signal might have an envelope similar to the oscilloscope display shown at A. The RF amplitude (current or voltage) is on the vertical axis, and the time sweep is on the horizontal axis. After speech processing, the evelope pattern might look like the display at B. The average power of the processed signal has increased, but the PEP has not changed.

G4D03 How is the output PEP of a transmitter calculated if an oscilloscope is used to measure the transmitter's peak load voltage across a resistive load?
A. PEP = [(Vp)(Vp)] / (RL)
B. PEP = [(0.707 PEV)(0.707 PEV)] / RL
C. PEP = (Vp)(Vp)(RL)
D. PEP = [(1.414 PEV)(1.414 PEV)] / RL

B The power used to measure a transmitter's output is peak envelope power (PEP). This is the average power of the wave at a modulation peak. When determining PEP, RMS voltage is used rather than peak voltage. RMS voltage is 0.707 times peak envelope voltage. The power in a circuit is equal to the current times the voltage. If you don't know the current but do know the voltage and resistance, you can substitute voltage divided by resistance for the current because current is equal to voltage divided by resistance. Then the formula for power becomes voltage times voltage divided by resistance. The voltage to be used in this case is 0.707 PEV, which stands for peak envelope voltage. The formula now becomes 0.707 PEV times 0.707 PEV divided by the resistive load.

$$PEP = \frac{0.707 \times 0.707 \text{ PEV}}{R_L} = \frac{(0.707 \text{ PEV})^2}{R_L}$$

Amateur Radio Practices   121

G4D04 What is the output PEP from a transmitter if an oscilloscope measures 200 volts peak-to-peak across a 50-ohm resistor connected to the transmitter output?
- A. 100 watts
- B. 200 watts
- C. 400 watts
- D. 1000 watts

A   The power used to measure a transmitter's output is peak envelope power (PEP). This is the average power of the wave at a modulation peak. When determining PEP, RMS voltage is used rather than peak voltage. RMS voltage is 0.707 times peak envelope voltage. The power in a circuit is equal to the current times the voltage. If you don't know the current but do know the voltage and resistance, you can substitute voltage divided by resistance for the current because current is equal to voltage divided by resistance. Then the formula for power becomes voltage times voltage divided by resistance. The voltage to be used in this case is 0.707 PEV, which stands for peak envelope voltage. The formula now becomes 0.707 PEV times 0.707 PEV divided by the resistive load. The formula in this case works out to:

$$\text{PEP} = \frac{0.707 \text{ PEV} \times 0.707 \text{ PEV}}{R_L} = \frac{(0.707 \times 100 \text{ V}) \times (0.707 \times 100 \text{ V})}{50}$$

$$\text{PEP} = \frac{(0.707 \times 100 \text{ V})^2}{50} = \frac{5000 \text{ V}^2}{50 \text{ }\Omega} = 100 \text{ W}$$

G4D05 What is the output PEP from a transmitter if an oscilloscope measures 500 volts peak-to-peak across a 50-ohm resistor connected to the transmitter output?

A. 500 watts
B. 625 watts
C. 1250 watts
D. 2500 watts

B The power used to measure a transmitter's output is peak envelope power (PEP). This is the average power of the wave at a modulation peak. When determining PEP, RMS voltage is used rather than peak voltage. RMS voltage is 0.707 times peak envelope voltage. The power in a circuit is equal to the current times the voltage. If you don't know the current but do know the voltage and resistance, you can substitute voltage divided by resistance for the current because current is equal to voltage divided by resistance. Then the formula for power becomes voltage times voltage divided by resistance. The voltage to be used in this case is 0.707 PEV, which stands for peak envelope voltage. The formula now becomes 0.707 PEV times 0.707 PEV divided by the resistive load. The formula in this case works out to:

$$PEP = \frac{0.707 \text{ PEV} \times 0.707 \text{ PEV}}{R_L} = \frac{(0.707 \times 250 \text{ V}) \times (0.707 \times 250 \text{ V})}{50}$$

$$PEP = \frac{(0.707 \times 250 \text{ V})^2}{50} = \frac{31{,}240 \text{ V}^2}{50 \text{ }\Omega} = 625 \text{ W}$$

G4D06 What is the output PEP of an unmodulated carrier transmitter if an average-reading wattmeter connected to the transmitter output indicates 1060 watts?

A. 530 watts
B. 1060 watts
C. 1500 watts
D. 2120 watts

B If a carrier is not modulated, then peak envelope power equals average power.

G4D07 Which wires in a four-conductor line cord should be attached to fuses in a 240-VAC primary (single phase) power supply?
- A. Only the "hot" (black and red) wires
- B. Only the "neutral" (white) wire
- C. Only the ground (bare) wire
- D. All wires

A   Since it is critically important that the neutral (white) wire always have continuity to ground, a fuse should never be installed on the white wire. On the other hand, the hot black and red wires should always have a fuse.

G4D08 What size wire is normally used on a 15-ampere, 120-VAC household lighting circuit?
- A. AWG number 14
- B. AWG number 16
- C. AWG number 18
- D. AWG number 22

A   When considering wire sizes, it is helpful to remember that bigger numbered wires can carry less current and smaller numbered wires can carry more current. A number 14 wire can carry 15 amps (numbers 14 and 15 are close to each other, making it easy to remember). On the other hand, number 12 wire can carry 20 amps.

G4D09 What size wire is normally used on a 20-ampere, 120-VAC household appliance circuit?
- A. AWG number 20
- B. AWG number 16
- C. AWG number 14
- D. AWG number 12

D   When considering wire sizes, it is helpful to remember that bigger numbered wires can carry less current and smaller numbered wires can carry more current. A number 12 wire can carry 20 amps. (You may find it helpful to remember the 2 in 12 and the 2 in 20 go together.) On the other hand, number 14 wire can carry 15 amps (numbers 14 and 15 are close to each other, making it easy to remember).

G4D10 What maximum size fuse or circuit breaker should be used in a household appliance circuit using AWG number 12 wiring?
A. 100 amperes
B. 60 amperes
C. 30 amperes
D. 20 amperes

D When considering wire sizes and fuse or circuit breaker ratings, it is helpful to remember that bigger numbered wires can carry less current and smaller numbered wires can carry more current. A number 12 wire can carry 20 amps. (You may find it helpful to remember the 2 in 12 and the 2 in 20 go together.) On the other hand, number 14 wire can carry 15 amps (numbers 14 and 15 are close to each other, making it easy to remember). The fuse or circuit breaker rating should always match the current carrying ability of the wire used for that circuit.

G4D11 What maximum size fuse or circuit breaker should be used in a household appliance circuit using AWG number 14 wiring?
A. 15 amperes
B. 20 amperes
C. 30 amperes
D. 60 amperes

A When considering wire sizes, it is helpful to remember that bigger numbered wires can carry less current and smaller numbered wires can carry more current. A number 14 wire carries 15 amps (numbers 14 and 15 are close to each other, making it easy to remember). On the other hand, number 12 wire carries 20 amps. The fuse or circuit breaker rating should always match the current carrying ability of the wire used for that circuit.

G4E Common connectors used in amateur stations: types; when to use; fastening methods; precautions when using; HF mobile radio installations; emergency power systems; generators; battery storage devices and charging sources including solar; wind generation

G4E01 Which of the following connectors is NOT designed for RF transmission lines?
A. PL-259
B. Type N
C. BNC
D. DB-25

D The PL-259 connector is sometimes called the UHF connector although it is generally considered to have too much loss for use at UHF. Type N and BNC connectors are also designed for use with coaxial cable RF transmission lines. A DB-25 connector is the 25-pin connector used as a serial port connector on many computers.

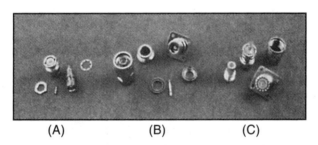

(A)   (B)   (C)

This photo shows some common coaxial-cable connectors. Part A shows a BNC connector pair. Many hand-held radios use BNC connectors. They are popular when a weatherproof connector is needed for RG-58 sized cable. Part B shows a pair of N connectors. These are often used for UHF equipment because of their low loss. Type N connectors provide a weatherproof connector for RG-8 sized cables. Part C shows a PL-259 coaxial connector and its mating SO-239 chassis connector. Most HF equipment uses these connectors. They are designed for use with RG-8 sized cables, although reducer bushings are available for smaller-diameter cables such as RG-58 and RG-59. Although the PL-259 is sometimes called a "UHF connector," it is seldom used on UHF equipment because it is considered to have too much loss for UHF.

G4E02 When installing a power plug on a line cord, which of the following should you do?

    A. Twist the wire strands neatly and fasten them so they don't cause a short circuit
    B. Observe the correct wire color conventions for plug terminals
    C. Use proper grounding techniques
    D. All of these choices

D   The reason you need to twist the wires tightly and wrap them securely around the proper screw terminals is that loose wires can come in contact with other terminals and cause a short circuit. What is the proper color coding? The black or red (hot) lead should connect to the brass-colored screw terminal. The white (neutral) lead should connect to the silver-colored terminal. The green or bare lead is the ground lead and should connect to the green-colored terminal or the terminal that connects to the longer, round pin of the plug.

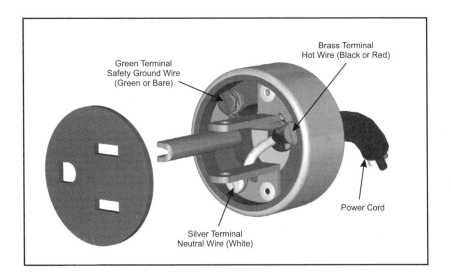

G4E03 Which of the following power connections would be the best for a 100-watt HF mobile installation?
  A. A direct, fused connection to the battery using heavy gauge wire
  B. A connection to the fuse-protected accessory terminal strip or distribution panel
  C. A connection to the cigarette lighter
  D. A direct connection to the alternator or generator

A When you are making the power connections for your 100-watt HF radio for mobile operation, you should connect both wires directly to the battery terminals. Remember that both leads should have fuses, placed as close to the battery as possible.

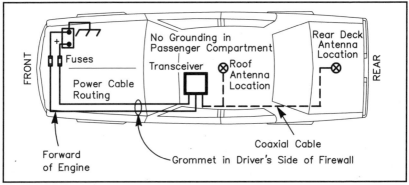

This drawing shows a typical mobile-transceiver-installation wiring diagram.

G4E04 Why is it best NOT to draw the DC power for a 100-watt HF transceiver from an automobile's cigarette lighter socket?
  A. The socket is not wired with an RF-shielded power cable
  B. The DC polarity of the socket is reversed from the polarity of modern HF transceivers
  C. The power from the socket is never adequately filtered for HF transceiver operation
  D. The socket's wiring may not be adequate for the current being drawn by the transceiver

D The cigarette lighter wiring was adequate for your low-power handheld radio but probably is not adequate for a full-power HF rig drawing 20 amps when transmitting. When you are making the power connections for your 100-watt HF radio for mobile operation, you should connect both wires directly to the battery terminals. Remember that both leads should have fuses, placed as close to the battery as possible.

G4E05 Which of the following most limits the effectiveness of an HF mobile transceiver operating in the 75-meter band?
  A. The vehicle's electrical system wiring
  B. The wire gauge of the DC power line to the transceiver
  C. The HF mobile antenna system
  D. The rating of the vehicle's alternator or generator

C   HF mobile antenna systems are the most limiting factor for effective operation of your station. Placing even an 8-foot vertical antenna on top of a small car makes a dangerously tall system. If you have a larger car, this kind of antenna is almost out of the question. Some type of inductive loading to shorten the length would be required. For operation on the 75-meter band this will make an inefficient antenna system.

G4E06 Which of the following is true of both a permanent or temporary emergency generator installation?
  A. The generator should be located in a well ventilated area
  B. The installation should be grounded
  C. Extra fuel supplies, especially gasoline, should not be stored in an inhabited area
  D. All of these choices

D   Carbon monoxide and other exhaust fumes can accumulate in your garage, basement or other confined living area, so ventilation is very important. You can ensure proper grounding for your generator through a connection to your electrical system ground. Store the fuel for your generator in an approved container or in an uninhabited area. Do not store fuel in the generator's fuel tank. Use this fuel in your car and get a new supply each month.

G4E07 Which of the following is true of a lead-acid storage battery as it is being charged?
  A. It tends to cool off
  B. It gives off explosive oxygen gas
  C. It gives off explosive hydrogen gas
  D. It takes in oxygen from the surrounding air

C   You need to keep your lead-acid battery charged at all times for use in an emergency. A by-product of charging your battery is the release of hydrogen gas which can explode if ignited by a spark. (Have you ever seen pictures of the airship Hindenberg?) A well-ventilated area is essential.

G4E08 What is the name of the process by which sunlight is directly changed into electricity?
   A. Photovoltaic conversion
   B. Photosensitive conduction
   C. Photosynthesis
   D. Photocoupling

A   Natural sources of energy are becoming more economical and more practical as a way to power your station. Photovoltaic conversion converts sunlight into electricity.

G4E09 What is the approximate open-circuit voltage from a modern, well illuminated photovoltaic cell?
   A. 0.02 VDC
   B. 0.2 VDC
   C. 1.38 VDC
   D. 0.5 VDC

D   Each photovoltaic cell produces about 0.5 volt in full sunlight if there is no load connected to the cell. The size or surface area of the cell determines the maximum current that the cell can supply.

G4E10 What determines the proper size solar panel to use in a solar-powered battery-charging circuit?
   A. The panel's voltage rating and maximum output current
   B. The amount of voltage available per square inch of panel
   C. The panel's open-circuit current
   D. The panel's short-circuit voltage

A   Solar panels are rated according to output voltage and maximum output current. When setting up your solar electric system, you need to keep this in mind. After the initial expense of setting up this system, the sun will provide a ready supply of free energy. You will need storage batteries for nighttime or for cloudy conditions.

G4E11 What is the biggest disadvantage to using wind power as the primary source of power for an emergency station?

- A. The conversion efficiency from mechanical energy to electrical energy is less that 2 percent
- B. The voltage and current ratings of such systems are not compatible with amateur equipment
- C. A large electrical storage system is needed to supply power when the wind is not blowing
- D. All of these choices are correct

C   Wind-powered systems, while complex in design and construction, provide free energy after the initial cost. Just like solar-powered systems, wind systems need to have a storage battery to supply electricity when the wind is not blowing.

Subelement G5

Electrical Principles

Your General class exam will have 2 questions taken from the 2 groups of questions in this Electrical Principles subelement; G5A and G5B.

G5A Impedance, including matching; resistance, including ohm; reactance; inductance; capacitance; and metric divisions of these values

G5A01 What is impedance?
  A. The electric charge stored by a capacitor
  B. The opposition to the flow of AC in a circuit containing only capacitance
  C. The opposition to the flow of AC in a circuit
  D. The force of repulsion between one electric field and another with the same charge

C Impedance is the opposition to the flow of current in an ac circuit. Capacitive reactance, inductive reactance, and resistance are all types of impedance.

G5A02 What is reactance?
  A. Opposition to DC caused by resistors
  B. Opposition to AC caused by inductors and capacitors
  C. A property of ideal resistors in AC circuits
  D. A large spark produced at switch contacts when an inductor is de-energized

B The impedance (opposition to flow of ac current) caused by inductors and capacitors in an ac circuit is referred to as reactance.

G5A03 In an inductor, what causes opposition to the flow of AC?
  A. Resistance
  B. Reluctance
  C. Admittance
  D. Reactance

D The impedance (opposition to flow of current) in an ac circuit caused by an inductor is referred to as inductive reactance.

G5A04 In a capacitor, what causes opposition to the flow of AC?
- A. Resistance
- B. Reluctance
- C. Reactance
- D. Admittance

C The impedance (opposition to flow of current) in an ac circuit caused by a capacitor is referred to as capacitive reactance.

G5A05 How does a coil react to AC?
- A. As the frequency of the applied AC increases, the reactance decreases
- B. As the amplitude of the applied AC increases, the reactance increases
- C. As the amplitude of the applied AC increases, the reactance decreases
- D. As the frequency of the applied AC increases, the reactance increases

D A coil in an ac circuit is referred to as an inductor. The impedance (opposition to flow of current) caused by the coil is referred to as inductive reactance. This reactance increases as the frequency of the ac current increases.

G5A06 How does a capacitor react to AC?
- A. As the frequency of the applied AC increases, the reactance decreases
- B. As the frequency of the applied AC increases, the reactance increases
- C. As the amplitude of the applied AC increases, the reactance increases
- D. As the amplitude of the applied AC increases, the reactance decreases

A The impedance (opposition to flow of current) in an ac circuit caused by a capacitor is referred to as capacitive reactance. This reactance decreases as the frequency in the ac current increases.

G5A07 When will a power source deliver maximum output to the load?
  A. When the impedance of the load is equal to the impedance of the source
  B. When the load resistance is infinite
  C. When the power-supply fuse rating equals the primary winding current
  D. When air wound transformers are used instead of iron-core transformers

A   A power source will deliver maximum power to a load when the impedance of the load is equal to the impedance of the source. When the impedances are not matched, the load will reflect power back to the source, setting up a standing wave.

G5A08 What happens when the impedance of an electrical load is equal to the internal impedance of the power source?
  A. The source delivers minimum power to the load
  B. The electrical load is shorted
  C. No current can flow through the circuit
  D. The source delivers maximum power to the load

D   A power source will deliver maximum power to a load when the impedance of the load is equal to the impedance of the source. When the impedances are not matched, the load will reflect power back to the source, setting up a standing wave.

G5A09 Why is impedance matching important?
  A. So the source can deliver maximum power to the load
  B. So the load will draw minimum power from the source
  C. To ensure that there is less resistance than reactance in the circuit
  D. To ensure that the resistance and reactance in the circuit are equal

A   A power source will deliver maximum power to a load when the impedance of the load is equal to the impedance of the source. When the impedances are not matched, the load will reflect power back to the source, setting up a standing wave.

Electrical Principles

G5A10 What unit is used to measure reactance?
   A. Mho
   B. Ohm
   C. Ampere
   D. Siemens

B The ohm is the unit used to measure any opposition to the flow of current. In an ac circuit, this opposition is referred to as impedance, and includes both reactance and resistance.

G5A11 What unit is used to measure impedance?
   A. Volt
   B. Ohm
   C. Ampere
   D. Watt

B The ohm is the unit used to measure any opposition to the flow of current. In an ac circuit, this opposition is referred to as impedance, and includes both reactance and resistance.

G5B Decibel; Ohm's Law; current and voltage dividers; electrical power calculations and series and parallel components; transformers (either voltage or impedance); sine wave root-mean-square (RMS) value

G5B01 A two-times increase in power results in a change of how many dB?
   A. 1 dB higher
   B. 3 dB higher
   C. 6 dB higher
   D. 12 dB higher

B The decibel scale is a logarithmic scale in which a two-times increase in power is represented by 3 dB. The mathematical formula for the decibel scale for power is:

$$dB = 10 \times \log_{10} \frac{(P_2)}{(P_1)}$$

where:

P_1 = reference power and P_2 = power being compared to the reference value.

$$dB = 10 \times \log_{10} \frac{(2)}{(1)} = 10 \times \log_{10}(2) = 10 \times 0.3 = 3 \text{ dB}$$

Some Common Decibel Values and Power Ratio Equivalents

dB	P_2/P_1	
20	100	(10^2)
10	10	(10^1)
6	4	
3	2	
0	1	
−3	0.5	
−6	0.25	
−10	0.1	(10^{-1})
−20	0.01	(10^{-2})

G5B02 In a parallel circuit with a voltage source and several branch resistors, how is the total current related to the current in the branch resistors?
    A. It equals the average of the branch current through each resistor
    B. It equals the sum of the branch current through each resistor
    C. It decreases as more parallel resistors are added to the circuit
    D. It is the sum of each resistor's voltage drop multiplied by the total number of resistors

B   In a circuit with several parallel resistors, the total current is equal to the sum of the current through each resistor.

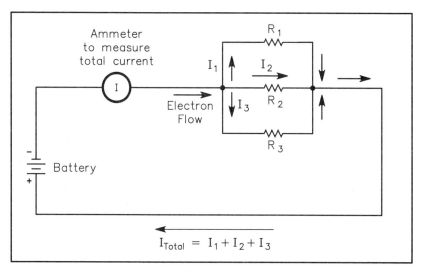

The sum of the current flowing into a junction point (node) in a circuit must be equal to the current flowing out of the junction point (node). This principle is called Kirchhoff's Current Law, named for Gustav Kirchhoff, the German scientist who discovered it.

Electrical Principles   137

G5B03 How many watts of electrical power are used if 400 VDC is supplied to an 800-ohm load?

    A. 0.5 watts
    B. 200 watts
    C. 400 watts
    D. 320,000 watts

B   Since $P = I \times E$ and $I = \dfrac{E}{R}$, the power in a circuit also can be expressed as $P = \dfrac{E \times E}{R}$. In this case:

$$P = \frac{E \times E}{R} = \frac{400 \times 400}{800} = \frac{160,000}{800} = 200 \text{ W}$$

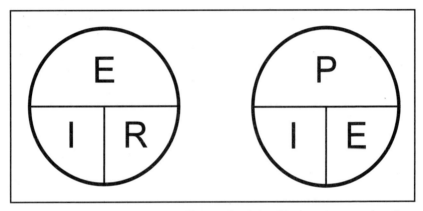

The "Ohm's Law Circle" and the "Power Circle" will help you remember the equations that include voltage, current, resistance and power. Cover the letter representing the unknown quantity to find an equation to calculate that quantity. If you cover the I in the Ohm's Law Circle, you are left with E/R. If you cover the P in the Power Circle, you are left with I × E. Combining these terms, you can write the equation to calculate power when you know the voltage and the resistance.

G5B04 How many watts of electrical power are used by a 12-VDC light bulb that draws 0.2 amperes?
- A. 60 watts
- B. 24 watts
- C. 6 watts
- D. 2.4 watts

D   Use the Power Circle drawing to find the equation to calculate power. The power in a circuit is equal to the volts times the current:

$P = I \times E = 0.2 \times 12 = 2.4$ W

G5B05 How many watts are being dissipated when 7.0 milliamperes flow through 1.25 kilohms?
- A. Approximately 61 milliwatts
- B. Approximately 39 milliwatts
- C. Approximately 11 milliwatts
- D. Approximately 9 milliwatts

A   Use the Ohm's Law Circle and Power Circle drawings to find the equations to calculate power. Since $P = I \times E$ and $E = R \times I$, the power in a circuit can also be expressed as $P = I \times I \times R$.

$P = I \times I \times R = 0.007 \times 0.007 \times 1250 = 0.06125$ W $= 61.25$ mW

Note: Remember that 7 milliamperes is equal to 0.007 ampere and 0.06125 watt is equal to approximately 61 milliwatts.

G5B06 What is the voltage across a 500-turn secondary winding in a transformer if the 2250-turn primary is connected to 120 VAC?

  A. 2370 volts
  B. 540 volts
  C. 26.7 volts
  D. 5.9 volts

C The voltage in the secondary winding of a transformer is equal to the voltage in the primary winding times the ratio of windings in the secondary to the primary.

$$E_S = E_P \times \frac{N_S}{N_P}$$

If the 2250-turn primary is connected to 120 volts ac, the voltage across a 500-turn secondary winding in the transformer is 26.7 volts:

$$E_S = E_P \times \frac{N_S}{N_P} = 120\,V \times \frac{500}{2250} = 26.7\,V$$

G5B07 What is the turns ratio of a transformer to match an audio amplifier having a 600-ohm output impedance to a speaker having a 4-ohm impedance?

  A. 12.2 to 1
  B. 24.4 to 1
  C. 150 to 1
  D. 300 to 1

A The turns ratio required to match impedances of circuits is equal to the square root of the impedance ratio. The turns ratio of a transformer to match an audio amplifier having a 600-ohm output impedance to a speaker having a 4-ohm impedance is 12.2 to 1. Take the square root of 600 / 4 (or the square root of 150).

$$\text{Turns Ratio} = \sqrt{\frac{600}{4}} = \sqrt{150} = 12.2$$

G5B08 What is the impedance of a speaker that requires a transformer with a turns ratio of 24 to 1 to match an audio amplifier having an output impedance of 2000 ohms?
  A. 576 ohms
  B. 83.3 ohms
  C. 7.0 ohms
  D. 3.5 ohms

D If you know the turns ratio and want to know the impedance ratio, square the turns ratio. In this case, the impedance ratio is 576 / 1 (24 × 24 = 576). Divide the higher impedance by the impedance ratio to find the lower impedance. In this case, $\frac{2000}{576}$ gives a speaker impedance of 3.5 ohms. In practice, you would use a 4-ohm speaker.

$$\frac{2000}{576} = 3.5$$

G5B09 A DC voltage equal to what value of an applied sine-wave AC voltage would produce the same amount of heat over time in a resistive element?
  A. The peak-to-peak value
  B. The RMS value
  C. The average value
  D. The peak value

B RMS refers to root-mean-square. RMS is a way of expressing an ac voltage that is equivalent to the dc voltage of the same value. (Both voltages will produce the same amount of heat in a heating element.)

G5B10 What is the peak-to-peak voltage of a sine wave that has an RMS voltage of 120 volts?
  A. 84.8 volts
  B. 169.7 volts
  C. 204.8 volts
  D. 339.4 volts

D If you know the RMS voltage and want to know the peak value, multiply the RMS by the square root of 2 (which is 1.414). If you want to know peak-to-peak, double the result. In this case, 120 × 1.414 × 2 = 339.4 V.

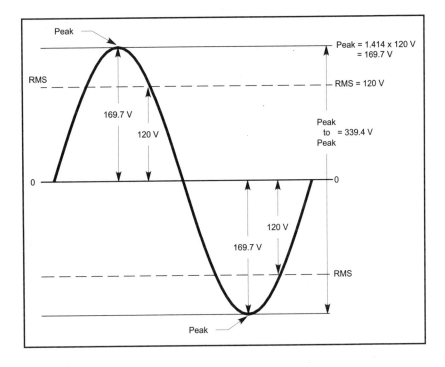

G5B11 A sine wave of 17 volts peak is equivalent to how many volts RMS?

    A. 8.5 volts
    B. 12 volts
    C. 24 volts
    D. 34 volts

B   If you know the peak voltage, you can find the RMS value by multiplying the peak voltage by 0.707 (which is the same as dividing by the square root of 2). 17 × 0.707 = 12 V.

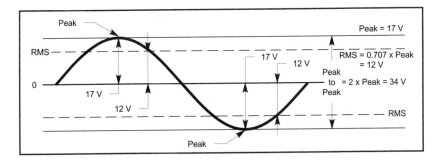

Subelement G6

Circuit Components

There will be 1 question from the Circuit Components subelement on your exam. That question will be taken from the questions pinted in this chapter.

G6A Resistors; capacitors; inductors; rectifiers and transistors; etc.

G6A01 If a carbon resistor's temperature is increased, what will happen to the resistance?
A. It will increase by 20% for every 10 degrees centigrade
B. It will stay the same
C. It will change depending on the resistor's temperature coefficient rating
D. It will become time dependent

C The resistance value of almost any substance changes when heated. The amount of change depends on the resistor's temperature coefficient rating.

G6A02 What type of capacitor is often used in power-supply circuits to filter the rectified AC?
A. Disc ceramic
B. Vacuum variable
C. Mica
D. Electrolytic

D   Because electrolytic capacitors can be made with the required high capacitance value in a small package, they are often used in power supply circuits to filter the rectified ac.

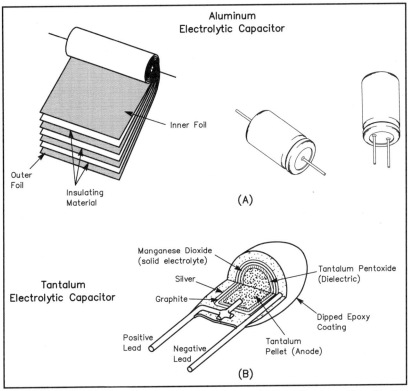

Part A shows the construction of an aluminum electrolytic capacitor. Part B shows the construction of a tantalum electrolytic capacitor.

G6A03 What function does a capacitor serve if it is used in a power-supply circuit to filter transient voltage spikes across the transformer's secondary winding?
- A. Clipper capacitor
- B. Trimmer capacitor
- C. Feedback capacitor
- D. Suppressor capacitor

D A suppressor capacitor takes its name from the fact that it suppresses otherwise damaging spikes and surges in a power-supply circuit.

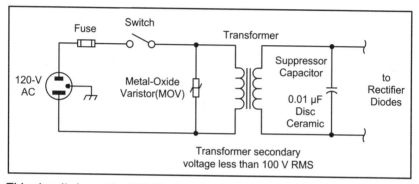

This circuit shows the 120-V ac input and transformer for a power supply. The MOV on the transformer primary suppresses transient spikes on the ac line voltage. The suppressor capacitor on the transformer secondary filters any transient voltage that comes through the transformer.

Circuit Components   145

G6A04 Where is the source of energy connected in a transformer?
- A. To the secondary winding
- B. To the primary winding
- C. To the core
- D. To the plates

B   When two coils are arranged so that a changing current in one coil induces a voltage in another, the combination is called a transformer. The source of energy is connected to the primary winding in a transformer. The load is attached to the secondary winding.

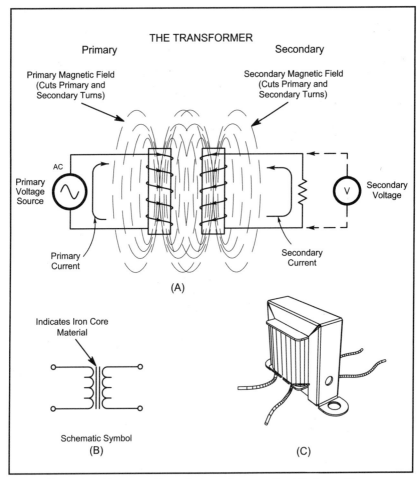

Part A illustrates the operation of a transformer. Part B shows the schematic symbol used to represent an iron-core transformer. Part C shows what a typical small transformer might look like.

146   Subelement G6

G6A05 If no load is attached to the secondary winding of a transformer, what is current in the primary winding called?
- A. Magnetizing current
- B. Direct current
- C. Excitation current
- D. Stabilizing current

A   When two coils are arranged so that a changing current in one coil induces a voltage in another, the combination is called a transformer. The source of energy is connected to the primary winding in a transformer. The load is attached to the secondary winding. If no load is attached to the secondary winding, the current in the primary is called the magnetizing current.

G6A06 What is the peak-inverse-voltage rating of a power-supply rectifier?
- A. The maximum transient voltage the rectifier will handle in the conducting direction
- B. 1.4 times the AC frequency
- C. The maximum voltage the rectifier will handle in the non-conducting direction
- D. 2.8 times the AC frequency

C   A power-supply rectifier diode is usually made from semiconductor material. Diodes allow current to flow only in one direction. If the voltage is strong enough, however, current can be forced to flow in the opposite direction. The voltage required to make this happen is the peak inverse voltage.

G6A07 What are the two major ratings that must not be exceeded for silicon-diode rectifiers used in power-supply circuits?
  A. Peak inverse voltage; average forward current
  B. Average power; average voltage
  C. Capacitive reactance; avalanche voltage
  D. Peak load impedance; peak voltage

A   A power-supply rectifier diode is usually made from semiconductor material. Diodes allow current to flow only in one direction. If the voltage is strong enough, however, current can be forced to flow in the opposite direction. The voltage required to make this happen is the peak inverse voltage. The average forward current is the maximum current that the circuit can take in the forward direction. Peak inverse voltage and average forward current are two major ratings that must not be exceeded for silicone diode rectifiers.

G6A08 What is the output waveform of an unfiltered full-wave rectifier connected to a resistive load?
  A. A series of pulses at twice the frequency of the AC input
  B. A series of pulses at the same frequency as the AC input
  C. A sine wave at half the frequency of the AC input
  D. A steady DC voltage

A   A full-wave rectifier changes alternating current with positive and negative half cycles to a fluctuating current with all positive pulses. Since the current in this case has not yet been filtered, it is a series of pulses at twice the frequency of the ac input.

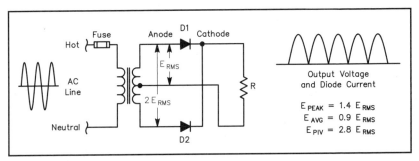

This circuit shows a simple full-wave rectifier with a load resistor but no filter. The waveform shown at the right represents the output voltage and diode current. The values of the peak output voltage, average output voltage and peak inverse voltage across the diodes are listed.

G6A09 A half-wave rectifier conducts during how many degrees of each cycle?

A. 90 degrees
B. 180 degrees
C. 270 degrees
D. 360 degrees

B   Since there are 360 degrees in a full cycle, a half-wave rectifier conducts during 180 degrees.

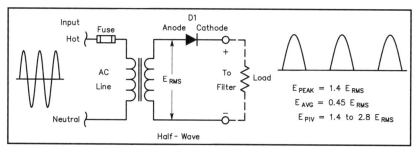

This circuit is a basic half-wave rectifier circuit, with a load resistor but no filter. The waveform shown at the right represents the output voltage and diode current. The values of the peak output voltage, average output voltage and peak inverse voltage across the diodes are listed.

G6A10 A full-wave rectifier conducts during how many degrees of each cycle?

A. 90 degrees
B. 180 degrees
C. 270 degrees
D. 360 degrees

D   Since there are 360 degrees in a full cycle, a full-wave rectifier conducts during all 360 degrees of each cycle. See the Figure for question G6A08.

Circuit Components   149

G6A11 When two or more diodes are connected in parallel to increase the current-handling capacity of a power supply, what is the purpose of the resistor connected in series with each diode?
- A. The resistors ensure that one diode doesn't take most of the current
- B. The resistors ensure the thermal stability of the power supply
- C. The resistors regulate the power supply output voltage
- D. The resistors act as swamping resistors in the circuit

A Two or more diodes are sometimes connected in parallel to increase the current-handling ability of the circuit. There should always be a resistor connected in series with each diode when this technique is used. Without a resistor in series with each diode, one diode may take most of the current, and that could destroy the diode. The resistor value should be chosen to provide a few tenths of a volt drop at the expected forward current.

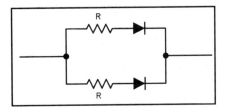

Use equalizing resistors when you connect diodes in parallel to increase the forward current-handling capability. The resistor value is chosen to provide a few tenths of a volt drop at the expected forward current.

Subelement G7

Practical Circuits

There will be 1 question from the Practical Circuits subelement on your exam. That question will be taken from the questions printed in this chapter.

G7A Power supplies and filters; single-sideband transmitters and receivers

G7A01 What safety feature does a power-supply bleeder resistor provide?
- A. It improves voltage regulation
- B. It discharges the filter capacitors
- C. It removes shock hazards from the induction coils
- D. It eliminates ground-loop current

B A bleeder resistor is installed across a filter capacitor in order to discharge the capacitor when power is not supplied to the circuit. This is to minimize the risk of electrical shock.

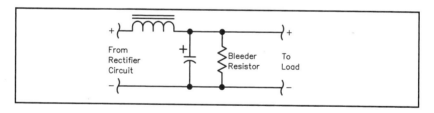

G7A02 Where is a power-supply bleeder resistor connected?
- A. Across the filter capacitor
- B. Across the power-supply input
- C. Between the transformer primary and secondary windings
- D. Across the inductor in the output filter

A A bleeder resistor is installed across a filter capacitor in order to discharge the capacitor when power is not supplied to the circuit. This is to minimize the risk of electrical shock. See the drawing with question G7A01.

Practical Circuits 151

G7A03 What components are used in a power-supply filter network?
   A. Diodes
   B. Transformers and transistors
   C. Quartz crystals
   D. Capacitors and inductors

D   A power supply filter network consists of capacitors and inductors used to smooth out the ripples in the rectified current. Capacitors oppose changes in voltage while the inductors oppose changes in current.

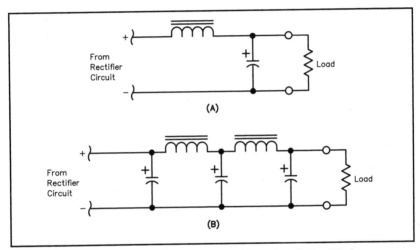

Part A shows a choke-input power-supply filter circuit. Part B shows a capacitor-input, multisection filter.

G7A04 What should be the minimum peak-inverse-voltage rating of the rectifier in a full-wave power supply?
   A. One-quarter the normal output voltage of the power supply
   B. Half the normal output voltage of the power supply
   C. Equal to the normal output voltage of the power supply
   D. Double the normal peak output voltage of the power supply

D   A power-supply rectifier diode is usually made from semiconductor material. Diodes allow current to flow only in one direction. If the voltage is strong enough, however, current can be forced to flow in the opposite direction. The voltage required to make this happen is the peak inverse voltage. It is good practice to make sure that the peak inverse voltage rating of the rectifier in a full-wave power supply is double the normal peak output voltage of the power supply.

G7A05 What should be the minimum peak-inverse-voltage rating of the rectifier in a half-wave power supply?
- A. One-quarter to one-half the normal peak output voltage of the power supply
- B. Half the normal output voltage of the power supply
- C. Equal to the normal output voltage of the power supply
- D. One to two times the normal peak output voltage of the power supply

D A power-supply rectifier diode is usually made from semiconductor material. Diodes allow current to flow only in one direction. If the voltage is strong enough, however, current can be forced to flow in the opposite direction. The voltage required to make this happen is the peak inverse voltage. It is good practice to make sure that the peak inverse voltage rating of the rectifier in a half-wave power supply is equal to one to two times the normal peak output voltage of the power supply.

G7A06 What should be the impedance of a low-pass filter as compared to the impedance of the transmission line into which it is inserted?
- A. Substantially higher
- B. About the same
- C. Substantially lower
- D. Twice the transmission line impedance

B It is wise to keep all elements of the feed line antenna system at the same impedance. This ensures that the maximum power is transferred from the transmitter to the antenna. A low-pass filter, designed to be installed at the transmitter output should have the same impedance as the transmission line.

G7A07 In a typical single-sideband phone transmitter, what circuit processes signals from the balanced modulator and sends signals to the mixer?
   A. Carrier oscillator
   B. Filter
   C. IF amplifier
   D. RF amplifier

B   In a single-sideband transmitter, the modulating audio is added to the RF signal in the balanced modulator. The balanced modulator also balances out or cancels the original carrier signal leaving a double-sideband, suppressed-carrier signal. This signal is sent to the filter, which filters out one of the sidebands, leaving a single-sideband signal.

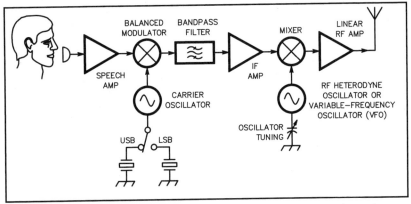

This block diagram shows a basic single-sideband, suppressed-carrier (SSB) transmitter.

G7A08 In a single-sideband phone transmitter, what circuit processes signals from the carrier oscillator and the speech amplifier and sends signals to the filter?
   A. Mixer
   B. Detector
   C. IF amplifier
   D. Balanced modulator

D   In a single-sideband transmitter, the modulating audio is added to the RF signal from the carrier oscillator in the balanced modulator. The balanced modulator also balances out or cancels the original carrier signal leaving a double-sideband, suppressed-carrier signal. This signal is sent to the filter, which filters out one of the sidebands, leaving a single-sideband signal. See the diagram with question G7A07.

154   Subelement G7

G7A09 In a single-sideband phone superheterodyne receiver, what circuit processes signals from the RF amplifier and the local oscillator and sends signals to the IF filter?
  A. Balanced modulator
  B. IF amplifier
  C. Mixer
  D. Detector

C   In a superheterodyne receiver, the mixer combines signals from the RF amplifier and the local oscillator and sends those signals to the filter, which passes the desired range of frequencies while rejecting the signals above and below the desired band.

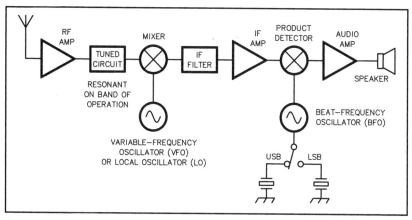

This block diagram shows a simple superheterodyne SSB receiver.

G7A10 In a single-sideband phone superheterodyne receiver, what circuit processes signals from the IF amplifier and the BFO and sends signals to the AF amplifier?
  A. RF oscillator
  B. IF filter
  C. Balanced modulator
  D. Detector

D   In a superheterodyne receiver, the detector circuit processes the signals from the intermediate frequency (IF) filter and the beat frequency oscillator (BFO) and passes the signal on to the audio frequency (AF) amplifier. The detector takes the information from an incoming radio signal and converts it to audio. See the drawing with question G7A09.

Practical Circuits   155

G7A11 In a single-sideband phone superheterodyne receiver, what circuit processes signals from the IF filter and sends signals to the detector?
- A. RF oscillator
- B. IF amplifier
- C. Mixer
- D. BFO

B   In a superheterodyne receiver, the mixer combines signals from the RF amplifier and the local oscillator and sends those signals to the intermediate frequency (IF) filter, which passes the desired range of frequencies while rejecting the signals above and below the desired band. Signals from the IF filter are then sent to the IF amplifier before going on to the detector. See the drawing with question G7A09.

Subelement G8

Signals and Emissions

There will be 2 questions on your general class exam from the Signals and Emisions subelement. These questions will be taken from the 2 groups of questions labeled G8A and G8B printed in this chapter.

G8A Signal information; AM; FM; single and double sideband and carrier; bandwidth; modulation envelope; deviation; overmodulation

G8A01 What type of modulation system changes the amplitude of an RF wave for the purpose of conveying information?
A. Frequency modulation
B. Phase modulation
C. Amplitude-rectification modulation
D. Amplitude modulation

D There are several ways to change or modulate an RF wave for the purpose of conveying information. CW simply turns the signal on and off to convey information. Amplitude modulation changes the strength or amplitude of the wave while frequency modulation changes its frequency. Phase modulation conveys information by changing the phase of the wave.

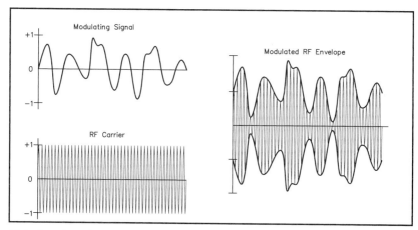

This drawing shows the relationship between the modulating audio waveform, the RF carrier and the resulting RF envelope in a double-sideband, full-carrier amplitude-modulated signal.

Signals and Emissions 157

G8A02 What type of modulation system changes the phase of an RF wave for the purpose of conveying information?
- A. Pulse modulation
- B. Phase modulation
- C. Phase-rectification modulation
- D. Amplitude modulation

B There are several ways to change or modulate an RF wave for the purpose of conveying information. CW simply turns the signal on and off to convey information. Amplitude modulation changes the strength or amplitude of the wave while frequency modulation changes its frequency. Phase modulation conveys information by changing the phase of the wave.

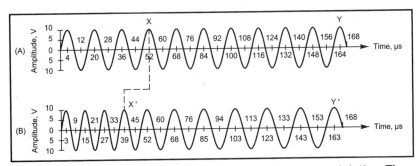

This drawing shows a graphical representation of phase modulation. The unmodulated wave is shown at A. Part B shows the modulated wave. After modulation, cycle X´ occurs earlier than cycle X did. All the cycles to the left of X´ are compressed, and to the right they are spread out.

G8A03 What type of modulation system changes the frequency of an RF wave for the purpose of conveying information?
- A. Phase-rectification modulation
- B. Frequency-rectification modulation
- C. Amplitude modulation
- D. Frequency modulation

D   There are several ways to change or modulate an RF wave for the purpose of conveying information. CW simply turns the signal on and off to convey information. Amplitude modulation changes the strength or amplitude of the wave while frequency modulation changes its frequency. Phase modulation conveys information by changing the phase of the wave.

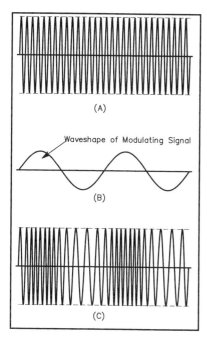

This drawing shows a graphical representation of frequency modulation. In the unmodulated carrier at Part A, each RF cycle takes the same amount of time to complete. When the modulating signal of Part B is applied, the carrier frequency is increased or decreased according to the amplitude and polarity of the modulating signal. Part C shows the modulated RF wave.

Signals and Emissions   159

G8A04 What emission is produced by a reactance modulator connected to an RF power amplifier?
   A. Multiplex modulation
   B. Phase modulation
   C. Amplitude modulation
   D. Pulse modulation

B   There are several ways to change or modulate an RF wave for the purpose of conveying information. CW simply turns the signal on and off to convey information. Amplitude modulation changes the strength or amplitude of the wave while frequency modulation changes its frequency. Phase modulation conveys information by changing the phase of the wave. In phase modulation, a reactance modulator is connected to an RF power amplifier in order to change the phase of the transmitted signal.

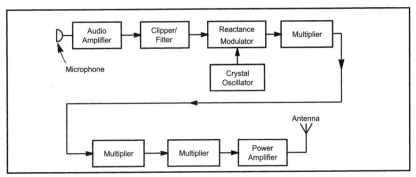

This block diagram shows a phase-modulated transmitter, which uses a reactance modulator. The transmitted signal is identical to the output from an FM transmitter.

G8A05 In what emission type does the instantaneous amplitude (envelope) of the RF signal vary in accordance with the modulating audio?
   A. Frequency shift keying
   B. Pulse modulation
   C. Frequency modulation
   D. Amplitude modulation

D   There are several ways to change or modulate an RF wave for the purpose of conveying information. CW simply turns the signal on and off to convey information. Amplitude modulation changes the strength or amplitude of the wave while frequency modulation changes its frequency. Phase modulation conveys information by changing the phase of the wave. In an AM transmission, at any given instant the amplitude or envelope of the RF signal changes according to the modulating audio signal. See the drawing with question G8A01.

G8A06 How much should the carrier be suppressed below peak output power in a properly designed single-sideband (SSB) transmitter?

A. No more than 20 dB
B. No more than 30 dB
C. At least 40 dB
D. At least 60 dB

C   In a single-sideband (SSB) amplitude-modulated transmitter, the RF carrier signal is reduced or suppressed, and one of the sidebands is removed. Suppression of the carrier allows more power to be put into the sideband. In a properly designed SSB transmitter, the carrier should be suppressed at least 40 dB below the peak output power of the sideband.

G8A07 What is one advantage of carrier suppression in a double-sideband phone transmission?

A. Only half the bandwidth is required for the same information content
B. Greater modulation percentage is obtainable with lower distortion
C. More power can be put into the sidebands
D. Simpler equipment can be used to receive a double-sideband suppressed-carrier signal

C   In a double-sideband, suppressed-carrier, amplitude-modulated transmitter, the RF carrier signal is reduced or suppressed. The less energy that is used by the carrier the more power that can be put into the sidebands. This increases the efficiency of the system.

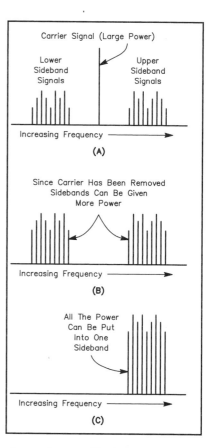

This drawing shows the frequency spectrum of an amplitude-modulated radio signal. Part A shows a double-sideband, full-carrier signal. When the carrier is removed, only the two sidebands are left, as shown in Part B. When one sideband is suppressed, as shown in Part C, the full transmitter power can be concentrated in one sideband.

Signals and Emissions   161

G8A08 Which popular phone emission uses the narrowest frequency bandwidth?
A. Single-sideband
B. Double-sideband
C. Phase-modulated
D. Frequency-modulated

A   In a single-sideband (SSB) amplitude-modulated transmitter, the RF carrier signal is reduced or suppressed, and one of the sidebands is removed. Suppression of the carrier allows more power to be put into the sideband. Because the carrier is suppressed and one sideband is filtered out, only enough frequency bandwidth is required to transmit a single sideband. This has the narrowest bandwidth of all the popular phone emissions.

G8A09 What happens to the signal of an overmodulated single-sideband or double-sideband phone transmitter?
A. It becomes louder with no other effects
B. It occupies less bandwidth with poor high-frequency response
C. It has higher fidelity and improved signal-to-noise ratio
D. It becomes distorted and occupies more bandwidth

D   When a signal is overmodulated it becomes distorted, making it difficult to understand. This can also cause splatter — a signal with excessive bandwidth that results in interference to nearby signals. Overmodulation is caused by having the microphone gain control adjusted too high or by overprocessing the signal with a speech processor. For proper adjustment, the microphone gain control should be adjusted so that there is a slight movement of the ALC meter on modulation peaks. An overmodulated signal is shown on an oscilloscope with a flattening off at the top of the modulation. This is referred to as flattopping.

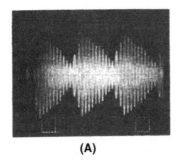
(A)

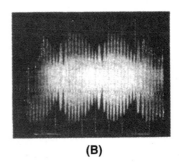
(B)

Part A shows the waveform of a properly adjusted SSB transmitter. Part B shows a severely clipped, distorted signal.

G8A10 How should the microphone gain control be adjusted on a single-sideband phone transmitter?
- A. For full deflection of the ALC meter on modulation peaks
- B. For slight movement of the ALC meter on modulation peaks
- C. For 100% frequency deviation on modulation peaks
- D. For a dip in plate current

B When a signal is overmodulated it becomes distorted, making it difficult to understand. This can also cause splatter — a signal with excessive bandwidth that results in interference to nearby signals. Overmodulation is caused by having the microphone gain control adjusted too high or by overprocessing the signal with a speech processor. For proper adjustment, the microphone gain control should be adjusted so that there is a slight movement of the ALC meter on modulation peaks. An overmodulated signal is shown on an oscilloscope with a flattening off at the top of the modulation. This is referred to as flattopping.

G8A11 What is meant by flattopping in a single-sideband phone transmission?
- A. Signal distortion caused by insufficient collector current
- B. The transmitter's automatic level control is properly adjusted
- C. Signal distortion caused by excessive drive
- D. The transmitter's carrier is properly suppressed

C When a signal is overmodulated it becomes distorted, making it difficult to understand. This can also cause splatter — a signal with excessive bandwidth that results in interference to nearby signals. Overmodulation is caused by having the microphone gain control adjusted too high or by overprocessing the signal with a speech processor. For proper adjustment, the microphone gain control should be adjusted so that there is a slight movement of the ALC meter on modulation peaks. An overmodulated signal is shown on an oscilloscope with a flattening off at the top of the modulation. This is referred to as flattopping. See the photo with question G8A09.

G8B Frequency mixing; multiplication; bandwidths; HF data communications

G8B01 What receiver stage combines a 14.25-MHz input signal with a 13.795-MHz oscillator signal to produce a 455-kHz intermediate frequency (IF) signal?
A. Mixer
B. BFO
C. VFO
D. Multiplier

A The mixer stage in a receiver combines the input signal with an oscillator signal to produce the sum and the difference of the two signals as well as the original two signals. Filters or mixer design will give one desired new signal, known as the intermediate frequency (IF) signal. For example, if a 13.795 MHz variable frequency oscillator (VFO) signal is mixed with a 14.25 MHz RF signal, it will produce a 455 kHz IF signal.

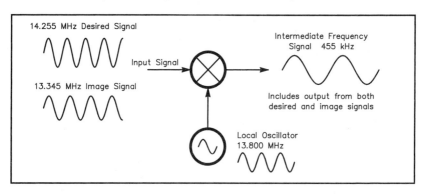

G8B02 If a receiver mixes a 13.800-MHz VFO with a 14.255-MHz received signal to produce a 455-kHz intermediate frequency (IF) signal, what type of interference will a 13.345-MHz signal produce in the receiver?

A. Local oscillator
B. Image response
C. Mixer interference
D. Intermediate interference

B   The mixer stage in a receiver combines the input signal with an oscillator signal to produce the sum and the difference of the two signals as well as the original two signals. Filters or mixer design will give one desired new signal, known as the intermediate frequency (IF) signal. In this case the input signal is 14.255 MHz and the oscillator signal is 13.800 MHz. The sum of these two signals is 28.055 and the difference, which is used as the intermediate frequency, is 0.455 MHz or 455 kHz. These signals are then sent to the filter to eliminate the unwanted frequencies leaving only the desired intermediate frequency of 455 kHz. A signal at 13.455 MHz, when subtracted from the 13.800 MHz oscillator signal will also produce the 455 kHz intermediate frequency. When, as in this case, an undesired input signal also produces a signal at the intermediate frequency, the resulting interference is called image response interference.

G8B03 What stage in a transmitter would change a 5.3-MHz input signal to 14.3 MHz?

A. A mixer
B. A beat frequency oscillator
C. A frequency multiplier
D. A linear translator

A   The mixer stage in a transmitter combines the local oscillator (LO) input signal with an RF signal at the intermediate frequency to produce the sum and the difference of the two signals as well as the original two signals. Filters or mixer design will give one desired new signal, which in this case is the desired RF output frequency. See the drawing with question G8B01.

G8B04 What is the name of the stage in a VHF FM transmitter that selects a harmonic of an HF signal to reach the desired operating frequency?
- A. Mixer
- B. Reactance modulator
- C. Preemphasis network
- D. Multiplier

D An FM transmitter often makes use of a device called a frequency multiplier. In a frequency multiplier, the input signal is used to modify the frequency of a high frequency (HF) oscillator. The multiplier selects one of the harmonics of the modulated signal, which is transmitted on the desired operating frequency.

G8B05 Why isn't frequency modulated (FM) phone used below 29.5 MHz?
- A. The transmitter efficiency for this mode is low
- B. Harmonics could not be attenuated to practical levels
- C. The bandwidth would exceed FCC limits
- D. The frequency stability would not be adequate

C At frequencies below 29.5 MHz, the FCC allowable bandwidths are narrower. The HF bands are relatively narrow, and would not be able to accommodate as many wide bandwidth signals.

G8B06 What is the total bandwidth of an FM-phone transmission having a 5-kHz deviation and a 3-kHz modulating frequency?

A. 3 kHz
B. 5 kHz
C. 8 kHz
D. 16 kHz

D To determine bandwidth of an FM-phone transmission, use the following formula:

$$Bw = 2 \times (D + M)$$

Where:

BW = bandwidth

D = frequency deviation (the instantaneous change in frequency for a given signal)

M = maximum modulating audio frequency

The total bandwidth of an FM-phone transmission having a 5 kHz deviation and a 3 kHz modulating frequency would be:

$$2 \times (5 \text{ kHz} + 3 \text{ kHz}) = 16 \text{ kHz}$$

G8B07 What is the frequency deviation for a 12.21-MHz reactance-modulated oscillator in a 5-kHz deviation, 146.52-MHz FM-phone transmitter?

A. 41.67 Hz
B. 416.7 Hz
C. 5 kHz
D. 12 kHz

B An FM transmitter often makes use of a device called a frequency multiplier. The HF oscillator signal is modulated to produce a signal that is amplified, and the frequency multiplier selects a harmonic of the signal for transmission on the desired frequency. The HF signal and amount of deviation are both multiplied. This is why you must set the oscillator deviation so the final deviation is what you want. To determine the oscillator frequency deviation, divide the output frequency by the oscillator frequency to determine the multiplication factor of the transmitter. Then divide the output deviation by the multiplication factor. The frequency deviation for a 12.21 MHz reactance-modulated oscillator in a 5 kHz deviation, 146.52 MHz FM-phone transmitter is 416.7 Hz.

$$\frac{146.52 \text{ MHz}}{12.21 \text{ MHz}} = 12$$

Then,

$$\frac{5{,}000 \text{ Hz}}{12} = 416.7 \text{ Hz}$$

G8B08 How is frequency shift related to keying speed in an FSK signal?
- A. The frequency shift in hertz must be at least four times the keying speed in WPM
- B. The frequency shift must not exceed 15 Hz per WPM of keying speed
- C. Greater keying speeds require greater frequency shifts
- D. Greater keying speeds require smaller frequency shifts

C   When high keying speeds are used, the receive demodulator tends to get confused if the frequency shifts are too close together. As a result, greater keying speeds require greater frequency shifts to allow proper reception.

G8B09 What do RTTY, Morse code, AMTOR and packet communications have in common?
- A. They are multipath communications
- B. They are digital communications
- C. They are analog communications
- D. They are only for emergency communications

B   RTTY, Morse code, AMTOR and packet radio are all communications methods that use a digital code to transfer information.

G8B10 What is the duty cycle required of a transmitter when sending Mode B (FEC) AMTOR?
- A. 50%
- B. 75%
- C. 100%
- D. 125%

C   Mode B AMTOR is a mode that sends information continuously without waiting for an acknowledgement. Consequently, the duty cycle required of a transmitter when sending Mode B AMTOR is 100%.

G8B11 In what segment of the 20-meter band are most AMTOR operations found?
- A. At the bottom of the slow-scan TV segment, near 14.230 MHz
- B. At the top of the SSB phone segment, near 14.325 MHz
- C. In the middle of the CW segment, near 14.100 MHz
- D. At the bottom of the RTTY segment, near 14.075 MHz

D   In order to minimize interference, concerned amateurs follow an agreement referred to as a band plan. The band plan specifies the location of various emissions on the band. On the 20-meter band, most of the AMTOR activity will be at the bottom of the RTTY segment, near 14.075 MHz.

Subelement G9

Antennas and Feed-Lines

Your General class exam will include 4 questions from the Antennas and Feed Lines subelement. These questions will be taken from the 4 groups of questions labeled G9A through G9D printed in this chapter.

G9A Yagi antennas - physical dimensions; impedance matching; radiation patterns; directivity and major lobes

G9A01 How can the SWR bandwidth of a parasitic beam antenna be increased?
A. Use larger diameter elements
B. Use closer element spacing
C. Use traps on the elements
D. Use tapered-diameter elements

A Using larger diameter elements can increase the SWR bandwidth of a parasitic beam antenna. The exact length of the elements becomes less critical when larger diameter elements are used.

G9A02 Approximately how long is the driven element of a Yagi antenna for 14.0 MHz?
A. 17 feet
B. 33 feet
C. 35 feet
D. 66 feet

B The driven element of the Yagi antenna should be $1/2$ wavelength. To find this length in feet, use the formula:

$$1/2 \text{ Wavelength (in feet)} = \frac{468}{f \text{ (in MHz)}} = \frac{468}{14.0 \text{ MHz}} = 33.4 \text{ feet}$$

The driven element of a Yagi antenna for 14.0 MHz should be approximately 33 feet.

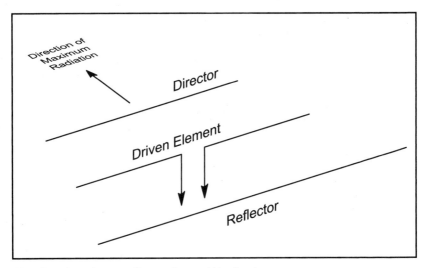

This drawing shows a three-element Yagi antenna.

G9A03 Approximately how long is the director element of a Yagi antenna for 21.1 MHz?
A. 42 feet
B. 21 feet
C. 17 feet
D. 10.5 feet

B The director element of a Yagi antenna should be 95% of the length of the driven element. See the drawing with question G9A02

First, find the ½-wavelength dipole or driven element length.

$$1/2 \text{ Wavelength (in feet)} = \frac{468}{f \text{ (in MHz)}} = \frac{468}{21.1 \text{ MHz}} = 22.3 \text{ feet}$$

Then, multiply this length by 0.95 to find the director-element length.

Director length (in feet) = 22.3 feet × 0.95 = 21.2 feet

The director element of a Yagi antenna for 21.1 MHz should be approximately 21 feet long.

G9A04 Approximately how long is the reflector element of a Yagi antenna for 28.1 MHz?
   A. 8.75 feet
   B. 16.6 feet
   C. 17.5 feet
   D. 35 feet

C The reflector element of a Yagi antenna should be 105% of the driven element. See the drawing with question G9A02.

First, find the ½-wavelength dipole or driven element length.

$$1/2 \text{ Wavelength (in feet)} = \frac{468}{f \text{ (in MHz)}} = \frac{468}{28.1 \text{ MHz}} = 16.7 \text{ feet}$$

Then, multiply this length by 1.05 to find the reflector-element length.

Reflector length (in feet) = 16.7 feet × 1.05 = 17.5 feet

The reflector element of a Yagi antenna for 28.1 MHz should be approximately 17.5 feet.

G9A05 Which statement about a three-element Yagi antenna is true?
   A. The reflector is normally the shortest parasitic element
   B. The director is normally the shortest parasitic element
   C. The driven element is the longest parasitic element
   D. Low feed-point impedance increases bandwidth

B In a Yagi, the director is 95% the length of the driven element and is placed at the front of the driven element in the direction the signal is being sent. The reflector element is 105% of the length of the driven element and is placed to the rear. The director is normally the shortest element of a Yagi antenna. See the drawing with question G9A02.

G9A06 What is one effect of increasing the boom length and adding directors to a Yagi antenna?
   A. Gain increases
   B. SWR increases
   C. Weight decreases
   D. Wind load decreases

A As the boom length of a Yagi is increased and more elements are added, the directivity or gain of the antenna increases. Directivity has an advantage in that it directs the signal in the intended direction more than in other directions, thus minimizing interference.

G9A07 Why is a Yagi antenna often used for radio communications on the 20-meter band?
- A. It provides excellent omnidirectional coverage in the horizontal plane
- B. It is smaller, less expensive and easier to erect than a dipole or vertical antenna
- C. It helps reduce interference from other stations off to the side or behind
- D. It provides the highest possible angle of radiation for the HF bands

C   A Yagi antenna provides gain or directivity. Directivity has an advantage in that it directs the signal in the intended direction more than in other directions, thus minimizing interference from stations in other directions. This is one important reason for using a Yagi antenna for HF operation, such as on the 20-meter band.

G9A08 What does "antenna front-to-back ratio" mean in reference to a Yagi antenna?
- A. The number of directors versus the number of reflectors
- B. The relative position of the driven element with respect to the reflectors and directors
- C. The power radiated in the major radiation lobe compared to the power radiated in exactly the opposite direction
- D. The power radiated in the major radiation lobe compared to the power radiated 90 degrees away from that direction

C   Using a directional antenna helps reduce interference in that it sends the signal in the intended direction, rather than off to the side or behind. Most of the radiated signal is sent in the desired direction. This is called the major lobe of radiation. A much smaller amount of the signal is radiated in other directions. If you measure the power radiated in the major lobe (or in the desired direction) and compare that with the power radiated in a direction exactly opposite to that, you have a measure of the antenna "front-to-back ratio."

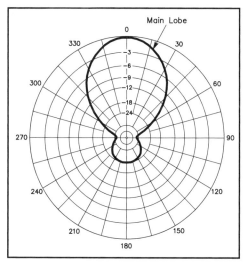

This diagram represents the directive pattern for a typical three-element Yagi antenna. Note that this pattern is essentially unidirectional, with most of the radiation in the direction of the main lobe. There is also a small *minor lobe* at 180° from the direction of the main lobe, however. You can read the front-to-back-ratio on this type of graph by finding the strength of the minor lobe off the back of the antenna. For this antenna, the front-to-back-ratio is 24 dB.

G9A09 What is the "main lobe" of a Yagi antenna radiation pattern?
- A. The direction of least radiation from the antenna
- B. The point of maximum current in a radiating antenna element
- C. The direction of maximum radiated field strength from the antenna
- D. The maximum voltage standing wave point on a radiating element

C   A Yagi antenna radiates most of the signal in the desired direction. This is called the major lobe of radiation. A much smaller amount of the signal is radiated in other directions. See the drawing with question G9A08.

G9A10 What is a good way to get maximum performance from a Yagi antenna?
- A. Optimize the lengths and spacing of the elements
- B. Use RG-58 feed-line
- C. Use a reactance bridge to measure the antenna performance from each direction around the antenna
- D. Avoid using towers higher than 30 feet above the ground

A   Using a directional antenna helps reduce interference in that it sends the signal in the intended direction, rather than off to the side or behind. Computer analysis shows that small changes in element lengths and spacing along the antenna boom can significantly alter the antenna performance. You can get the best performance from a Yagi antenna you are building if you optimize the element lengths and spacing along the boom.

Antennas and Feed Lines   173

G9A11 Which of the following is NOT a Yagi antenna design variable that should be considered to optimize the forward gain, front-to-rear gain ratio and SWR bandwidth?
A. The physical length of the boom
B. The number of elements on the boom
C. The spacing of each element along the boom
D. The polarization of the antenna elements

D There are a number of changes you can make to a Yagi antenna design to obtain the best performance. The physical length of the boom, the number of elements along the boom and the spacing of those elements will all affect the forward gain, front-to-rear gain ratio and the SWR bandwidth. Unfortunately, there is no single setting that will give the maximum performance for all these quantities at the same time. The polarization of the antenna elements (whether they are arranged horizontally or vertically) will not generally affect the antenna design, however.

G9B Loop antennas - physical dimensions; impedance matching; radiation patterns; directivity and major lobes

G9B01 Approximately how long is each side of a cubical-quad antenna driven element for 21.4 MHz?
A. 1.17 feet
B. 11.7 feet
C. 47 feet
D. 469 feet

B A cubical-quad antenna has a driven element that is a square-shaped loop. It also uses a square-shaped loop reflector and sometimes one or more directors. The entire driven element of a cubical-quad antenna is approximately a full wavelength. To determine the appropriate size of the driven element, in feet, use the formula:

$$1 \text{ Wavelength Driven Element (in feet)} = \frac{1005}{f \text{ (in MHz)}} = \frac{1005}{21.4 \text{ MHz}} = 46.96 \text{ feet}$$

To find the length of just one side, divide by 4.

$$\text{Length of 1 side of Driven Element (in feet)} = \frac{46.96 \text{ feet}}{4} = 11.7 \text{ feet}$$

Each side of a cubical-quad antenna driven element for 21.4 MHz should be approximately 11.7 feet.

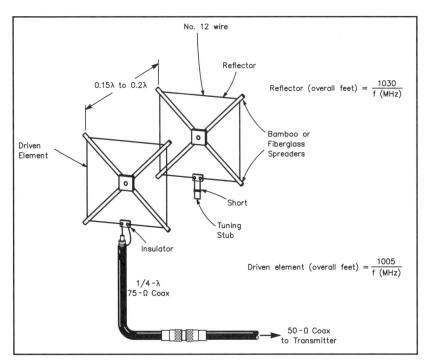

This drawing shows the constuction of a cubical-quad antenna. The lengths of the driven element and reflector element are given by the equations shown in the drawing. To find the length of one side, divide these total lengths by 4.

G9B02 Approximately how long is each side of a cubical-quad antenna driven element for 14.3 MHz?
A. 17.6 feet
B. 23.4 feet
C. 70.3 feet
D. 175 feet

A A cubical-quad antenna has a driven element that is a square-shaped loop. It also uses a square-shaped loop reflector and sometimes one or more directors. See the drawing with question G9B01. The entire driven element of a cubical-quad antenna is approximately a full wavelength. To determine the appropriate size of the driven element, use the formula:

$$1 \text{ Wavelength Driven Element (in feet)} = \frac{1005}{f \text{ (in MHz)}} = \frac{1005}{14.3 \text{ MHz}} = 70.3 \text{ feet}$$

To find the length of just one side, divide by 4.

$$\text{Length of 1 side of Driven Element (in feet)} = \frac{70.3 \text{ feet}}{4} = 17.6 \text{ feet}$$

Each side of a cubical-quad antenna driven element for 14.3 MHz should be approximately 17.6 feet.

G9B03 Approximately how long is each side of a cubical-quad antenna reflector element for 29.6 MHz?
A. 8.23 feet
B. 8.7 feet
C. 9.7 feet
D. 34.8 feet

B A cubical-quad antenna has a driven element that is a square-shaped loop. It also uses a square-shaped loop reflector and sometimes one or more directors. See the drawing with question G9B01. The entire reflector element of a cubical-quad antenna is slightly more than a full wavelength. To determine the appropriate size of the reflector element, use the formula:

$$1 \text{ Wavelength Reflector (in feet)} = \frac{1030}{f \text{ (in MHz)}} = \frac{1030}{29.6 \text{ MHz}} = 34.8 \text{ feet}$$

To find the length of just one side, divide by 4.

$$\text{Length of 1 side of Reflector (in feet)} = \frac{34.8 \text{ feet}}{4} = 8.7 \text{ feet}$$

Each side of a cubical-quad antenna reflector element for 29.6 MHz should be approximately 8.7 feet.

G9B04 Approximately how long is each leg of a symmetrical delta-loop antenna driven element for 28.7 MHz?
- A. 8.75 feet
- B. 11.7 feet
- C. 23.4 feet
- D. 35 feet

B A delta-loop antenna has a driven element that is a triangle-shaped loop. It also uses a triangle-shaped loop reflector and sometimes one or more directors. The entire driven element of a delta-loop antenna is approximately a full wavelength. To determine the appropriate size of the driven element, use the formula:

$$1 \text{ Wavelength Driven Element (in feet)} = \frac{1005}{f \text{ (in MHz)}} = \frac{1005}{28.7 \text{ MHz}} = 35.0 \text{ feet}$$

To find the length of just one side, divide by 3.

$$\text{Length of 1 side of Driven Element (in feet)} = \frac{35.0 \text{ feet}}{3} = 11.67 \text{ feet}$$

Each leg of a symmetrical delta-loop antenna driven element for 28.7 MHz should be approximately 11.7 feet.

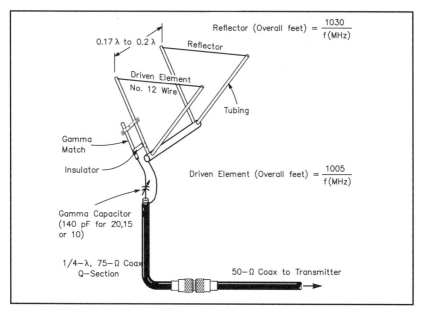

This drawing shows the construction of a delta-loop antenna. The lengths of the driven element and reflector element are given by the equations shown in the drawing. To find the length of one side, divide these total lengths by 3.

Antennas and Feed Lines

G9B05 Approximately how long is each leg of a symmetrical delta-loop antenna driven element for 24.9 MHz?

    A. 10.99 feet
    B. 12.95 feet
    C. 13.45 feet
    D. 40.36 feet

C   A delta-loop antenna has a driven element that is a triangle-shaped loop. It also uses a triangle-shaped loop reflector and sometimes one or more directors. See the drawing with question G9B04. The entire driven element of a delta-loop antenna is approximately a full wavelength. To determine the appropriate size of the driven element, use the formula:

$$1 \text{ Wavelength Driven Element (in feet)} = \frac{1005}{f \text{ (in MHz)}} = \frac{1005}{24.9 \text{ MHz}} = 40.4 \text{ feet}$$

To find the length of just one side, divide by 3.

$$\text{Length of 1 side of Driven Element (in feet)} = \frac{40.4 \text{ feet}}{3} = 13.47 \text{ feet}$$

Each leg of a symmetrical delta-loop antenna driven element for 24.9 MHz should be approximately 13.45 feet.

G9B06 Approximately how long is each leg of a symmetrical delta-loop antenna reflector element for 14.1 MHz?

    A. 18.26 feet
    B. 23.76 feet
    C. 24.35 feet
    D. 73.05 feet

C   A delta-loop antenna has a driven element that is a triangle-shaped loop. It also uses a triangle-shaped loop reflector and sometimes one or more directors. See the drawing with question G9B04. The entire reflector element of a delta-loop antenna is slightly more than a full wavelength. To determine the appropriate size of the reflector element, use the formula:

$$1 \text{ Wavelength Reflector (in feet)} = \frac{1030}{f \text{ (in MHz)}} = \frac{1030}{14.1 \text{ MHz}} = 73.0 \text{ feet}$$

To find the length of just one side, divide by 3.

$$\text{Length of 1 side of Reflector (in feet)} = \frac{73.0 \text{ feet}}{3} = 24.33 \text{ feet}$$

Each leg of a symmetrical delta-loop antenna reflector element for 14.1 MHz should be approximately 24.35 feet.

G9B07 Which statement about two-element delta loops and quad antennas is true?
- A. They compare favorably with a three-element Yagi
- B. They perform poorly above HF
- C. They perform very well only at HF
- D. They are effective only when constructed using insulated wire

A   A two-element quad or delta loop has about the same gain as a three-element Yagi.

G9B08 Compared to a dipole antenna, what are the directional radiation characteristics of a cubical-quad antenna?
- A. The quad has more directivity in the horizontal plane but less directivity in the vertical plane
- B. The quad has less directivity in the horizontal plane but more directivity in the vertical plane
- C. The quad has more directivity in both horizontal and vertical planes
- D. The quad has less directivity in both horizontal and vertical planes

C   The purpose of using a cubical-quad antenna is to obtain gain or directivity. Directivity has an advantage in that it directs the signal in the intended direction more than in other directions, thus minimizing interference. A cubical-quad antenna has more gain than a dipole antenna in both the horizonal and vertical planes. Again, this means more of the radio signal energy is aimed at the horizon.

G9B09 Moving the feed point of a multielement quad antenna from a side parallel to the ground to a side perpendicular to the ground will have what effect?
- A. It will significantly increase the antenna feed-point impedance
- B. It will significantly decrease the antenna feed-point impedance
- C. It will change the antenna polarization from vertical to horizontal
- D. It will change the antenna polarization from horizontal to vertical

D   By simply changing the feed point on a quad antenna, it is relatively easy to change the polarization. If you feed the antenna in the center of a side that is parallel to the ground, the antenna will have horizontal polarization. If you feed the antenna in the center of a vertical side, the antenna will have vertical polarization.

G9B10 What does the term "antenna front-to-back ratio" mean in reference to a delta-loop antenna?
  A. The number of directors versus the number of reflectors
  B. The relative position of the driven element with respect to the reflectors and directors
  C. The power radiated in the major radiation lobe compared to the power radiated in exactly the opposite direction
  D. The power radiated in the major radiation lobe compared to the power radiated 90 degrees away from that direction

C   Using a directional antenna helps reduce interference in that it sends the signal in the intended direction, rather than off to the side or behind. Most of the radiated signal is sent in the desired direction. This is called the major lobe of radiation. A much smaller amount of the signal is radiated in other directions. If you measure the power radiated in the major lobe (or in the desired direction) and compare that with the power radiated in a direction exactly opposite to that, you have a measure of the antenna "front-to-back ratio." A high front-to-back ratio is desirable in that it minimizes interference by sending more of the signal in the intended direction. See the drawing with question G9A08.

G9B11 What is the "main lobe" of a delta-loop antenna radiation pattern?
  A. The direction of least radiation from an antenna
  B. The point of maximum current in a radiating antenna element
  C. The direction of maximum radiated field strength from the antenna
  D. The maximum voltage standing wave point on a radiating element

C   Directional antennas, such as the delta-loop antenna, provide the advantage of directivity, sending more of the signal in the desired direction, thus minimizing interference. Most of the radiated signal is sent in the desired direction. This is called the major lobe of radiation. A much smaller amount of the signal is radiated in other directions. If you measure the power radiated in the major lobe (or in the desired direction) and compare that with the power radiated in a direction exactly opposite to that, you have a measure of the antenna "front-to-back ratio." A high front-to-back ratio is desirable in that it minimizes interference by sending more of the signal in the intended direction. See the drawing with question G9A08.

G9C Random wire antennas - physical dimensions; impedance matching; radiation patterns; directivity and major lobes; feed point impedance of 1/2-wavelength dipole and 1/4-wavelength vertical antennas

G9C01 What type of multiband transmitting antenna does NOT require a feed-line?
- A. A random-wire antenna
- B. A triband Yagi antenna
- C. A delta-loop antenna
- D. A Beverage antenna

A A random-length wire antenna is often referred to as a random-wire antenna. It consists simply of a wire. A random-length wire antenna can be of any length because an antenna tuner is used to match the impedances. It does not require a feed line because the single piece of wire serves as both a feed line and an antenna. It is considered to be a multiband antenna. You may experience RF feedback in your station when using a random-length wire antenna.

G9C02 What is one advantage of using a random-wire antenna?
- A. It is more efficient than any other kind of antenna
- B. It will keep RF energy out of your station
- C. It doesn't need an impedance matching network
- D. It is a multiband antenna

D A random-length wire antenna is often referred to as a random-wire antenna. It consists simply of a wire. A random-wire antenna can be of any length because an antenna tuner is used to match the impedances. It does not require a feed line because the single piece of wire serves as both a feed line and an antenna. It is considered to be a multiband antenna, because the antenna tuner will match the impedances for several frequency bands. You may experience RF feedback in your station when using a random-length wire antenna.

G9C03 What is one disadvantage of a random-wire antenna?
A. It must be longer than 1 wavelength
B. You may experience RF feedback in your station
C. It usually produces vertically polarized radiation
D. You must use an inverted-T matching network for multiband operation

B A random-length wire antenna is often referred to as a random-wire antenna. It consists simply of a wire. A random-wire antenna can be of any length because an antenna tuner is used to match the impedances. It does not require a feed line because the single piece of wire serves as both a feed line and an antenna. It is considered to be a multiband antenna, because the antenna tuner will match the impedances for several frequency bands. You may experience RF feedback in your station when using a random-length wire antenna.

G9C04 What is an advantage of downward sloping radials on a ground-plane antenna?
A. It lowers the radiation angle
B. It brings the feed-point impedance closer to 300 ohms
C. It increases the radiation angle
D. It brings the feed-point impedance closer to 50 ohms

D A ground-plane antenna has a $1/4$ wavelength vertical radiating element and four $1/4$ wavelength horizontal "radial" elements. You can change the impedance of a ground-plane antenna by changing the angle of the radials. Bending or sloping the radials downward to about a 45-degree angle will increase the impedance from approximately 35 ohms to approximately 50 ohms.

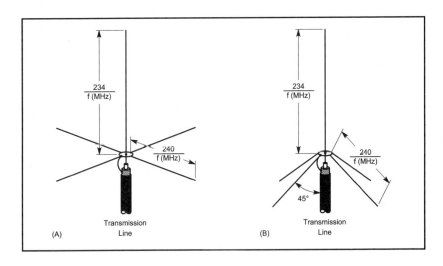

G9C05 What happens to the feed-point impedance of a ground-plane antenna when its radials are changed from horizontal to downward-sloping?

- A. It decreases
- B. It increases
- C. It stays the same
- D. It approaches zero

B   A ground-plane antenna has a ¹/₄ wavelength vertical radiating element and four ¹/₄ wavelength horizontal "radial" elements. You can change the impedance of a ground-plane antenna by changing the angle of the radials. Bending or sloping the radials downward to about a 45-degree angle will increase the impedance from approximately 35 ohms to approximately 50 ohms. See the drawing with question G9C04.

G9C06 What is the low-angle radiation pattern of an ideal half-wavelength dipole HF antenna installed a half-wavelength high, parallel to the earth?

- A. It is a figure-eight at right angles to the antenna
- B. It is a figure-eight off both ends of the antenna
- C. It is a circle (equal radiation in all directions)
- D. It is two smaller lobes on one side of the antenna, and one larger lobe on the other side

A   A ¹/₂-wavelength dipole antenna radiates its signals in a bi-directional fashion at a 90-degree angle to the antenna wires. This is called a "figure 8" radiation pattern. If the antenna is placed less than ¹/₂ wavelength above the ground, the ideal signal pattern will become distorted.

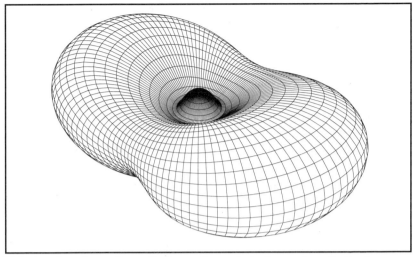

This drawing shows the radiation pattern of a half-wavelength dipole installed ½-wavelength above ground.

G9C07 How does antenna height affect the horizontal (azimuthal) radiation pattern of a horizontal dipole HF antenna?
A. If the antenna is too high, the pattern becomes unpredictable
B. If the antenna is less than one-half wavelength high, the azimuthal pattern is almost omnidirectional
C. Antenna height has no effect on the pattern
D. If the antenna is less than one-half wavelength high, radiation off the ends of the wire is eliminated

B A ½-wavelength dipole antenna radiates its signals in a bi-directional fashion at a 90-degree angle to the antenna wires. This is called a "figure 8" radiation pattern. If the antenna is placed less than ½ wavelength above the ground, the ideal signal pattern will become distorted. For heights of less than ½ wavelength, the antenna pattern becomes almost omnidirectional, sending signals nearly equally in all compass directions.

G9C08 If the horizontal radiation pattern of an antenna shows a major lobe at 0 degrees and a minor lobe at 180 degrees, how would you describe the radiation pattern of this antenna?
- A. Most of the signal would be radiated towards 180 degrees and a smaller amount would be radiated towards 0 degrees
- B. Almost no signal would be radiated towards 0 degrees and a small amount would be radiated towards 180 degrees
- C. Almost all the signal would be radiated equally towards 0 degrees and 180 degrees
- D. Most of the signal would be radiated towards 0 degrees and a smaller amount would be radiated towards 180 degrees

D If you use a field-strength meter to measure the radiated signal field intensity around an antenna, you can determine the general shape of the radiation pattern. If you find a major lobe at 0 degrees and a minor lobe at 180 degrees, most of the signal will be going in a direction of 0 degrees. Only a small amount of the radiated signal will be going in a direction of 180 degrees. See the drawing with question G9A08.

G9C09 If a slightly shorter parasitic element is placed 0.1 wavelength away and parallel to an HF dipole antenna mounted above ground, what effect will this have on the antenna's radiation pattern?
- A. The radiation pattern will not be affected
- B. A major lobe will develop in the horizontal plane, parallel to the two elements
- C. A major lobe will develop in the vertical plane, away from the ground
- D. A major lobe will develop in the horizontal plane, toward the parasitic element

D A $1/2$-wavelength dipole antenna radiates its signals in a bi-directional fashion at a 90-degree angle to the antenna wires. If the antenna is placed less than $1/2$ wavelength above the ground, the ideal signal pattern will become distorted. You can increase the directivity of a dipole antenna by using parasitic elements. This can be done by placing a slightly shorter parasitic element 0.1 wavelength away from the antenna in front of it and/or a slightly longer parasitic element 0.1 wavelength away from the antenna behind it. A major lobe of radiation will form along the plane of the elements, toward the element that is slightly shorter. See drawing on next page.

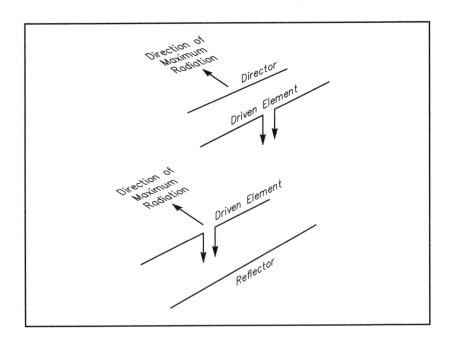

G9C10 If a slightly longer parasitic element is placed 0.1 wavelength away and parallel to an HF dipole antenna mounted above ground, what effect will this have on the antenna's radiation pattern?
- A. The radiation pattern will not be affected
- B. A major lobe will develop in the horizontal plane, away from the parasitic element, toward the dipole
- C. A major lobe will develop in the vertical plane, away from the ground
- D. A major lobe will develop in the horizontal plane, parallel to the two elements

B A ¹/₂-wavelength dipole antenna radiates its signals in a bi-directional fashion at a 90-degree angle to the antenna wires. If the antenna is placed less than ¹/₂ wavelength above the ground, the ideal signal pattern will become distorted. You can increase the directivity of a dipole antenna by using parasitic elements. This can be done by placing a slightly shorter parasitic element 0.1 wavelength away from the antenna in front of it and/or a slightly longer parasitic element 0.1 wavelength away from the antenna behind it. A major lobe of radiation will form along the plane of the elements, away from the element that is slightly longer. See the drawing with question G9C09.

G9C11 Where should the radial wires of a ground-mounted vertical antenna system be placed?
   A. As high as possible above the ground
   B. On the surface or buried a few inches below the ground
   C. Parallel to the antenna element
   D. At the top of the antenna

B   In most installations, ground conductivity is inadequate, so an artificial ground must be made from wires placed along the ground near the base of the antenna. These radials are usually 1/4 wavelength or longer. Depending on ground conductivity, 8, 16, 32 or more radials may be required to form an effective ground. The radial wires of a ground-mounted vertical antenna should be placed on the ground surface or buried a few inches below ground.

G9D Popular antenna feed-lines - characteristic impedance and impedance matching; SWR calculations

G9D01 Which of the following factors help determine the characteristic impedance of a parallel-conductor antenna feed-line?
   A. The distance between the centers of the conductors and the radius of the conductors
   B. The distance between the centers of the conductors and the length of the line
   C. The radius of the conductors and the frequency of the signal
   D. The frequency of the signal and the length of the line

A   The characteristic impedance of a parallel-conductor feed line depends on the distance between the conductor centers and the radius of the conductors.

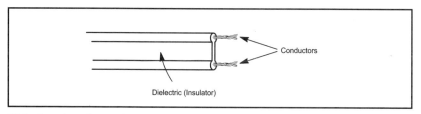

This drawing shows the construction of common 300-ohm twin lead.

Antennas and Feed Lines   187

G9D02 What is the typical characteristic impedance of coaxial cables used for antenna feed-lines at amateur stations?
   A. 25 and 30 ohms
   B. 50 and 75 ohms
   C. 80 and 100 ohms
   D. 500 and 750 ohms

   B   Common coaxial cables used as antenna feed lines have characteristic impedances of 50 or 75 ohms.

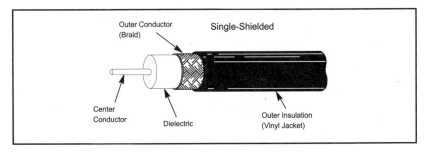

Coaxial cables consist of a conductor surrounded by insulation. The second conductor, called the shield, goes around the insulation. A plastic insulation goes around the entire cable.

G9D03 What is the characteristic impedance of flat-ribbon TV-type twin-lead?
   A. 50 ohms
   B. 75 ohms
   C. 100 ohms
   D. 300 ohms

   D   The flat ribbon type of feed line often used with TV antennas has a characteristic impedance of 300 ohms. This feed line is called twin lead. See the drawing with question G9D01.

G9D04 What is the typical cause of power being reflected back down an antenna feed-line?
   A. Operating an antenna at its resonant frequency
   B. Using more transmitter power than the antenna can handle
   C. A difference between feed line impedance and antenna feed-point impedance
   D. Feeding the antenna with unbalanced feed-line

C   Power reflected back from the antenna returns to the transmitter, which in turn reflects the power back towards the antenna. This creates a standing wave. The problem with high SWR really isn't the efficiency of the rig (power in / power out). When the transmitter and antenna impedances are not matched, less power is transferred to the antenna.

Modern solid-state transmitters usually have protection circuitry that reduces the power output (and input) in the presence of high SWR. (When the SWR is high there are high voltages present that can damage components.) In order to avoid this situation, the antenna feed-point impedance must be matched to the characteristic impedance of the feed line.

G9D05 What must be done to prevent standing waves of voltage and current on an antenna feed-line?
   A. The antenna feed point must be at DC ground potential
   B. The feed line must be cut to an odd number of electrical quarter-wavelengths long
   C. The feed line must be cut to an even number of physical half wavelengths long
   D. The antenna feed-point impedance must be matched to the characteristic impedance of the feed-line

D   Power reflected back from the antenna returns to the transmitter, which in turn reflects the power back towards the antenna. This creates a standing wave. The problem with high SWR really isn't the efficiency of the rig (power in / power out). When the transmitter and antenna impedances are not matched, less power is transferred to the antenna. Modern solid-state transmitters usually have protection circuitry that reduces the power output (and input) in the presence of high SWR. (When the SWR is high there are high voltages present that can damage components.) In order to avoid this situation, the antenna feed-point impedance must be matched to the characteristic impedance of the feed line.

G9D06 If a center-fed dipole antenna is fed by parallel-conductor feed-line, how would an inductively coupled matching network be used in the antenna system?
- A. It would not normally be used with parallel-conductor feed-lines
- B. It would be used to increase the SWR to an acceptable level
- C. It would be used to match the unbalanced transmitter output to the balanced parallel-conductor feed-line
- D. It would be used at the antenna feed point to tune out the radiation resistance

C Most transmitters are designed to accommodate unbalanced coaxial cable. In order to use a balanced parallel-conductor feed line, a matching network should be used. An inductively coupled matching network can be a good choice for this application because it will also match the unbalanced transmitter output to the balanced feed line.

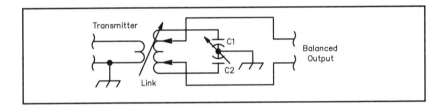

G9D07 If a 160-meter signal and a 2-meter signal pass through the same coaxial cable, how will the attenuation of the two signals compare?
- A. It will be greater at 2 meters
- B. It will be less at 2 meters
- C. It will be the same at both frequencies
- D. It will depend on the emission type in use

A Line loss is greater at higher frequencies. If you were to use the same type of coaxial cable feed line for your 160-meter antenna as for your 2-meter antenna, there would be much more loss at the 2 meter frequencies.

G9D08 In what values are RF feed line losses usually expressed?
- A. Bels/1000 ft
- B. dB/1000 ft
- C. Bels/100 ft
- D. dB/100 ft

D RF feed line loss is normally specified in decibels for each 100 feet of line.

G9D09 What standing-wave-ratio will result from the connection of a 50-ohm feed line to a resonant antenna having a 200-ohm feed-point impedance?

    A. 4:1
    B. 1:4
    C. 2:1
    D. 1:2

A   If a load on a feed line is purely resistive, the SWR can be calculated by dividing the line characteristic impedance by the load resistance or vice versa, whichever gives a value greater than one. 200 / 50 = 4:1 SWR.

G9D10 What standing-wave-ratio will result from the connection of a 50-ohm feed line to a resonant antenna having a 10-ohm feed-point impedance?

    A. 2:1
    B. 50:1
    C. 1:5
    D. 5:1

D   If a load on a feed line is purely resistive, the SWR can be calculated by dividing the line characteristic impedance by the load resistance or vice versa, whichever gives a value greater than one. 50 / 10 = 5:1 SWR.

G9D11 What standing-wave-ratio will result from the connection of a 50-ohm feed line to a resonant antenna having a 50-ohm feed-point impedance?

    A. 2:1
    B. 50:50
    C. 0:0
    D. 1:1

D   If a load on a feed line is purely resistive, the SWR can be calculated by dividing the line characteristic impedance by the load resistance or vice versa, whichever gives a value greater than one. 50 / 50 = 1:1 SWR.

Subelement G0

RF Safety

There will be 5 questions on your general class exam from the RF Safety subelement. Those 5 questions will be taken from the 5 groups of questions labeled G0A through G0E, printed in this chapter.

G0A RF Safety Principles

G0A01 Depending on the wavelength of the signal, the energy density of the RF field, and other factors, in what way can RF energy affect body tissue?
   A. It heats body tissue
   B. It causes radiation poisoning
   C. It causes the blood count to reach a dangerously low level
   D. It cools body tissue

A Body tissues that are subjected to very high levels of RF energy may suffer heat damage. These effects depend on the frequency of the energy, the power density of the RF field that strikes the body, and even on factors such as the polarization of the wave. The thermal effects of RF energy should not be a major concern for most radio amateurs because of the relatively low RF power we normally use and the intermittent nature of most amateur transmissions. It is rare for amateurs to be subjected to RF fields strong enough to produce thermal effects unless they are fairly close to an energized antenna or unshielded power amplifier.

G0A02 Which property is NOT important in estimating RF energy's effect on body tissue?
   A. Its duty cycle
   B. Its critical angle
   C. Its power density
   D. Its frequency

B The body's natural resonant frequencies affect how the body absorbs RF energy. For this reason, polarization, power density, and the frequency of the radio signal are all important in estimating the effects of RF energy on body tissue. The critical angle refers to an entirely different, unrelated concept. It is the maximum angle above the horizon at which a radio signal will be returned to the earth.

G0A03 Which of the following has the most direct effect on the exposure level of RF radiation?
A. The maximum usable frequency of the ionosphere
B. The frequency (or wavelength) of the energy
C. The environment near the transmitter
D. The distance from the antenna in the far field

B Body tissues that are subjected to very high levels of RF energy may suffer heat damage. These effects depend upon the frequency (or wavelength) of the energy, the power density of the RF field that strikes the body, and even on factors such as polarization of the wave.

G0A04 What unit of measurement best describes the biological effects of RF fields at frequencies used by amateur operators?
A. Electric field strength (V/m)
B. Magnetic field strength (A/m)
C. Specific absorption rate (W/kg)
D. Power density (W/cm^2)

C Specific Absorption Rate (SAR) is a term that describes the rate at which RF energy is absorbed into the human body tissue. This quantity, expressed in watts per kilogram (W/kg), best relates the RF exposure to the biological effects of RF fields on the human body at frequencies used by amateur operators.

G0A05 RF radiation in which of the following frequency ranges has the most effect on the human eyes?
A. The 3.5 MHz range
B. The 2 MHz range
C. The 50 MHz range
D. The 1270 MHz range

D Your head and structures such as your eyes will absorb energy in the 1270 MHz range and higher frequencies more readily than in the HF or UHF range.

G0A06 What does the term "athermal effects" of RF radiation mean?
- A. Biological effects from RF energy other than heating
- B. Chemical effects from RF energy on minerals and liquids
- C. A change in the phase of a signal resulting from the heating of an antenna
- D. Biological effects from RF energy in excess of the maximum permissible exposure level

A In addition to the heating effects of RF energy, scientists have also studied "athermal" or nonheating effects. This relates to the study of biological effects of RF other than heating of tissue. Some of the research into biological effects of RF radiation in recent years include studies that show that even fairly low levels of EMR can alter the body's circadian rhythms, affect the manner in which cancer-fighting T lymphocytes function in the immune system, and alter the nature of the electrical and chemical signals communicated through the cell membrane and between cells. Still, these studies do not show any effect of these low-level fields on the overall organism.

G0A07 At what frequencies does the human body absorb RF energy at a maximum rate?
- A. The high-frequency (3-30 MHz) range
- B. The very-high-frequency (30-300 MHz) range
- C. The ultra-high-frequency (300 MHz to 3 GHz) range
- D. The super-high-frequency (3 GHz to 30 GHz) range

B The human body absorbs RF energy at a maximum rate in the very-high-frequency range between 30 and 300 MHz. (This is why the lowest E-field exposure limit is in the 30 to 300 MHz range. The lowest H-field exposure levels occur at 100 to 300 MHz.)

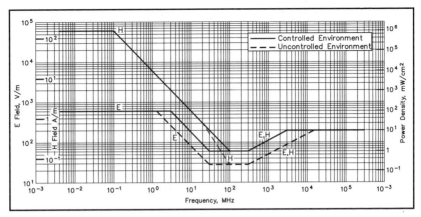

This graph shows the 1991 RF protection guidelines for body exposure of humans. It is known officially as the "IEEE Standard for Safety Levels with Respect to Human Exposure to Radio Frequency Electromagnetic Fields, 3 kHz to 300 GHz." (Note that the exposure levels set by this standard are not the same as the FCC permissible exposure limits.) This graph is necessarily complex because the standards differ not only for controlled and uncontrolled environments but also for electric fields (E fields) and magnetic fields (H fields).

G0A08 What does "time averaging" mean when it applies to RF radiation exposure?
    A. The average time of day when the exposure occurs
    B. The average time it takes RF radiation to have any long term effect on the body
    C. The total time of the exposure, e.g. 6 minutes or 30 minutes
    D. The total RF exposure averaged over a certain time

    D   Time averaging, when applied to RF radiation exposure, takes into account the total RF exposure averaged over either a 6-minute or a 30-minute exposure time. Time averaging adjusts the transmit/receive time ratio during normal amateur communications. It takes into account that the body cools itself after a time of reduced or no radiation exposure.

G0A09 What guideline is used to determine whether or not a routine RF evaluation must be performed for an amateur station?
  A. If the transmitter's PEP is 50 watts or more, an evaluation must always be performed
  B. If the RF radiation from the antenna system falls within a controlled environment, an evaluation must be performed
  C. If the RF radiation from the antenna system falls within an uncontrolled environment, an evaluation must be performed
  D. If the transmitter's PEP and frequency are within certain limits given in Part 97, an evaluation must be performed

D FCC guidelines given in Part 97 of the regulations list the peak envelope power (PEP) and frequency limitations that determine if you need to perform an RF Environmental Evaluation.

Power Thresholds for Routine Evaluation of Amateur Radio Stations

Wavelength Band	Evaluation Required if Power* (watts) Exceeds:
MF	
160 m	500
HF	
80 m	500
75 m	500
40 m	500
30 m	425
20 m	225
17 m	125
15 m	100
12 m	75
10 m	50
VHF	
(All Bands)	50
UHF	
70 cm	70
33 cm	150
23 cm	200
13 cm	250
SHF	
(All Bands)	250
EHF	
(All Bands)	250
Repeater stations	
(All Bands)	*Non-building-mounted antennas*: height above ground level to lowest point of antenna < 10 m *and* power > 500 W ERP
	Building-mounted antennas: power > 500 W ERP

*Transmitter power = Peak-envelope-power input to antenna. For repeater stations *only*, power exclusion based on ERP (effective radiated power).

G0A10 If you perform a routine RF evaluation on your station and determine that its RF fields exceed the FCC's exposure limits in human-accessible areas, what are you required to do?
   A. Take action to prevent human exposure to the excessive RF fields
   B. File an Environmental Impact Statement (EIS-97) with the FCC
   C. Secure written permission from your neighbors to operate above the controlled MPE limits
   D. Nothing; simply keep the evaluation in your station records

A   Some of the things you can do to prevent human exposure to excessive RF radiation are to move your antennas farther away, restrict access to the areas where exposure would exceed the limits, or reduce power to reduce the field strengths in those areas.

G0A11 At a site with multiple transmitters operating at the same time, how is each transmitter included in the RF exposure site evaluation?
   A. Only the RF field of the most powerful transmitter need be considered
   B. The RF fields of all transmitters are multiplied together
   C. Transmitters that produce more than 5% of the maximum permissible power density exposure limit for that transmitter must be included
   D. Only the RF fields from any transmitters operating with high duty-cycle modes (greater than 50%) need to be considered

C   In a multi-transmitter environment, each transmitter operator may be jointly responsible, with all other site operators, for ensuring that the total RF exposure from the site does not exceed the MPE limits. Any transmitter that produces more than 5% of the total permissible exposure limit for that transmitter must be included in the site evaluation. (This is 5% of the permitted power density or 5% of the square of the E or H-field MPE limit. It is NOT 5% of the total exposure, which sometimes can be unknown.)

G0B RF Safety Rules and Guidelines

G0B01 What are the FCC's RF-safety rules designed to control?
A. The maximum RF radiated electric field strength
B. The maximum RF radiated magnetic field strength
C. The maximum permissible human exposure to all RF radiated fields
D. The maximum RF radiated power density

C The core of the requirements under the regulations is the MPE level. The specific actions that need to be taken by amateur operators is the requirement for some amateurs to perform a routine RF Environmental Evaluation for RF exposure.

G0B02 At a site with multiple transmitters, who must ensure that all FCC RF-safety regulations are met?
A. All licensees contributing more than 5% of the maximum permissible power density exposure for that transmitter are equally responsible
B. Only the licensee of the station producing the strongest RF field is responsible
C. All of the stations at the site are equally responsible, regardless of any station's contribution to the total RF field
D. Only the licensees of stations which are producing an RF field exceeding the maximum permissible exposure limit are responsible

A In a multi-transmitter environment, each transmitter operator mat be jointly responsible, with all other site operators, for ensuring that the total RF exposure from the site does not exceed the MPR limits. Any transmitter that produces more than 5% of the total permissible exposure limit what for that transmitter must be included in the site evaluation. (This is 5% of the permitted power density or 5% of the square of the E or H-field MPE limit. It is NOT 5% of the total exposure, which sometimes can be unknown.)

G0B03 What effect does duty cycle have when evaluating RF exposure?
    A. Low duty-cycle emissions permit greater short-term exposure levels
    B. High duty-cycle emissions permit greater short-term exposure levels
    C. The duty cycle is not considered when evaluating RF exposure
    D. Any duty cycle may be used as long as it is less than 100 percent

A   Since amateurs spend more time listening than transmitting, low duty cycles are common. Remember that duty cycle takes into account the reduced average transmitted power that results because the transmitter is not operating at full power continuously. This means greater short-term exposure levels can be permitted with low-duty-cycle emissions.

G0B04 What is the threshold power used to determine if an RF environmental evaluation is required when the operation takes place in the 15-meter band?
    A. 50 watts PEP
    B. 100 watts PEP
    C. 225 watts PEP
    D. 500 watts PEP

B   The power levels used to determine if an RF Environmental Evaluation is required vary with frequency. See the Table with question G0A09. When you compare the MPE levels for various frequencies with the evaluation threshold levels, you will see that the evaluation thresholds vary in much the same way as the MPE limits. Where higher exposure is permitted, the evaluation threshold is higher; where MPE is lower, so is the evaluation threshold. On the 15-meter band you can transmit with up to 100 W PEP without being required to perform an RF Environmental Evaluation.

G0B05 Why do the power levels used to determine if an RF environmental evaluation is required vary with frequency?
- A. Because amateur operators may use a variety of power levels
- B. Because Maximum Permissible Exposure (MPE) limits are frequency dependent
- C. Because provision must be made for signal loss due to propagation
- D. All of these choices are correct

B The power levels used to determine if an RF Environmental Evaluation is required vary with frequency. When you compare the MPE levels for various frequencies with the evaluation threshold levels, you will see that the evaluation thresholds vary in much the same way as the MPE limits. Where higher exposure is permitted, the evaluation threshold is higher; where MPE is lower, so is the evaluation threshold.

(From §1.1310) Limits for Maximum Permissible Exposure (MPE)
(A) Limits for Occupational/Controlled Exposure

Frequency Range (MHz)	Electric Field Strength (V/m)	Magnetic Field Strength (A/m)	Power Density (mW/cm²)	Averaging Time (minutes)
0.3-3.0	614	1.63	(100)*	6
3.0-30	1842/f	4.89/f	(900/f²)*	6
30-300	61.4	0.163	1.0	6
300-1500	—	—	f/300	6
1500-100,000	—	—	5	6

f = frequency in MHz
* = Plane-wave equivalent power density (see Note 1).

(B) Limits for General Population/Uncontrolled Exposure

Frequency Range (MHz)	Electric Field Strength (V/m)	Magnetic Field Strength (A/m)	Power Density (mW/cm2)	Averaging Time (minutes)
0.3-1.34	614	1.63	(100)*	30
1.34-30	824/f	2.19/f	(180/f2)*	30
30-300	27.5	0.073	0.2	30
300-1500	—	—	f/1500	30
1500-100,000	—	—	1.0	30

f = frequency in MHz
* = Plane-wave equivalent power density (See Note 1.)
Note 1: This means the equivalent far-field strength that would have the E or H-field component calculated or measured. It does not apply well in the near field of an antenna. The equivalent far-field power density can be found in the near or far field regions from the relationships:

$P_d = |E_{total}|^2 / 3770$ mW/cm² or from $P_d = |H_{total}|^2 \times 37.7$ mW/cm².

G0B06 What is the threshold power used to determine if an RF environmental evaluation is required when the operation takes place in the 10-meter band?

- A. 50 watts PEP
- B. 100 watts PEP
- C. 225 watts PEP
- D. 500 watts PEP

A The power levels used to determine if an RF Environmental Evaluation is required vary with frequency. See the Table with question G0A09. When you compare the MPE levels for various frequencies with the evaluation threshold levels, you will see that the evaluation thresholds vary in much the same way as the MPE limits. Where higher exposure is permitted, the evaluation threshold is higher; where MPE is lower, so is the evaluation threshold. On the 10-meter band, you can transmit with up to 50 W PEP without being required to perform an RF Environmental Evaluation.

G0B07 What is the threshold power used to determine if an RF environmental evaluation is required for transmissions in the amateur bands with frequencies less than 10 MHz?

- A. 50 watts PEP
- B. 100 watts PEP
- C. 225 watts PEP
- D. 500 watts PEP

D The power levels used to determine if an RF Environmental Evaluation is required vary with frequency. See the Table with question G0A09. When you compare the MPE levels for various frequencies with the evaluation threshold levels, you will see that the evaluation thresholds vary in much the same way as the MPE limits. Where higher exposure is permitted, the evaluation threshold is higher; where MPE is lower, so is the evaluation threshold. On the 30-meter band (10 MHz), you can transmit with up to 500 W PEP without being required to perform an RF Environmental Evaluation.

G0B08 What amateur frequency bands have the lowest power limits above which an RF environmental evaluation is required?
- A. All bands between 17 and 30 meters
- B. All bands between 10 and 15 meters
- C. All bands between 40 and 160 meters
- D. All bands between 1.25 and 10 meters

D See the Table with question G0A09. If you want to use 12 and 10 meters with a 100-watt transceiver, you either could perform an RF Environmental Evaluation for those two bands, or you could reduce power to 75 W PEP on 12 meters and 50 W PEP on 10 meters and forego the evaluation. Most VHF transceivers are rated at 50 W PEP output or less; stations using this power level on VHF (the 6, 2, and 1.25-meter bands) would not need to be evaluated. All stations must ensure that no one is exposed to more than the maximum permissible exposure (MPE) levels.

G0B09 What is the threshold power used to determine if an RF safety evaluation is required when the operation takes place in the 20-meter band?
- A. 50 watts PEP
- B. 100 watts PEP
- C. 225 watts PEP
- D. 500 watts PEP

C The power levels used to determine if an RF Environmental Evaluation is required vary with frequency. See the Table with question G0A09. When you compare the MPE levels for various frequencies with the evaluation threshold levels, you will see that the evaluation thresholds vary in much the same way as the MPE limits. Where higher exposure is permitted, the evaluation threshold is higher; where MPE is lower, so is the evaluation threshold. On the 20-meter band, you can transmit with up to 225 W PEP without being required to perform an RF Environmental Evaluation.

G0B10 This question has been withdrawn.

G0B11 Under what conditions would an RF environmental evaluation be required for an amateur repeater station where the transmitting antenna is mounted on a building?
    A. The repeater transmitter is activated for more than 6 minutes without 30 seconds pauses
    B. The height above ground to the lowest point of the antenna is less than 10 m and the radiated power from the antenna exceeds 50 W ERP
    C. The height above ground to the lowest point of the antenna is less than 2 m and the radiated power from the antenna exceeds 50 W ERP
    D. The radiated power from the antenna exceeds 500 W ERP

D   See the Table with question G0A09. A routine environmental evaluation is required for an amateur repeater station using an antenna that is mounted on a building if the radiated power from the antenna is more than 500 W effective radiated power (ERP).

G0C Routine Station Evaluation and Measurements (FCC Part 97 refers to RF Radiation Evaluation)

G0C01 If the free-space far-field strength of a 10-MHz dipole antenna measures 1.0 millivolts per meter at a distance of 5 wavelengths, what will the field strength measure at a distance of 10 wavelengths?
A. 0.10 millivolts per meter
B. 0.25 millivolts per meter
C. 0.50 millivolts per meter
D. 1.0 millivolts per meter

C The electric field strength decreases linearly with the distance from the antenna. When you are twice as far from the antenna, the electric field strength will be ½ as strong. When you are 4 times farther away the electric field strength will be ¼ as strong. We can write a simple equation based on this relationship to calculate the field strength at some new distance when the strength is known at one distance.

$$\text{Field Strength}_2 = \text{Field Strength}_1 \times \frac{\text{Distance}_1}{\text{Distance}_2}$$

$$\text{Field Strength}_2 = 1.0 \, \frac{\text{mV}}{\text{m}} \times \frac{5\lambda}{10\lambda}$$

$$\text{Field Strength}_2 = 1.0 \, \frac{\text{mV}}{\text{m}} \times \frac{1}{2}$$

$$\text{Field Strength}_2 = 0.50 \, \frac{\text{mV}}{\text{m}}$$

G0C02 If the free-space far-field strength of a 28-MHz Yagi antenna measures 4.0 millivolts per meter at a distance of 5 wavelengths, what will the field strength measure at a distance of 20 wavelengths?
A. 2.0 millivolts per meter
B. 1.0 millivolts per meter
C. 0.50 millivolts per meter
D. 0.25 millivolts per meter

B The electric field strength decreases linearly with the distance from the antenna. When you are twice as far from the antenna, the electric field strength will be ½ as strong. When you are 4 times farther away the electric field strength will be ¼ as strong. We can write a simple equation based on this relationship to calculate the field strength at some new distance when the strength is known at one distance.

$$\text{Field Strength}_2 = 4.0 \frac{mV}{m} \times \frac{5\lambda}{20\lambda}$$

$$\text{Field Strength}_2 = 4.0 \frac{mV}{m} \times \frac{1}{4}$$

$$\text{Field Strength}_2 = 1.0 \frac{mV}{m}$$

G0C03 If the free-space far-field strength of a 1.8-MHz dipole antenna measures 9 microvolts per meter at a distance of 4 wavelengths, what will the field strength measure at a distance of 12 wavelengths?
A. 3 microvolts per meter
B. 3.6 microvolts per meter
C. 4.8 microvolts per meter
D. 10 microvolts per meter

A The electric field strength decreases linearly with the distance from the antenna. When you are twice as far from the antenna, the electric field strength will be ½ as strong. When you are 4 times farther away the electric field strength will be ¼ as strong. We can write a simple equation based on this relationship to calculate the field strength at some new distance when the strength is known at one distance.

$$\text{Field Strength}_2 = 9.0 \frac{\mu V}{m} \times \frac{4\lambda}{12\lambda}$$

$$\text{Field Strength}_2 = 9.0 \frac{\mu V}{m} \times \frac{1}{3}$$

$$\text{Field Strength}_2 = 3.0 \frac{\mu V}{m}$$

G0C04 If the free-space far-field power density of a 18-MHz Yagi antenna measures 10 milliwatts per square meter at a distance of 3 wavelengths, what will it measure at a distance of 6 wavelengths?
A. 11 milliwatts per square meter
B. 5.0 milliwatts per square meter
C. 3.3 milliwatts per square meter
D. 2.5 milliwatts per square meter

D The power density decreases as the square of the distance from the antenna. When you are twice as far from the antenna, the electric field strength will be ¼ as strong. When you are 4 times farther away, the electric field strength will be $1/16$ as strong. We can write a simple equation based on this relationship to calculate the power density at some new distance when the power density is known at one distance.

$$\text{Power Density}_2 = \text{P. Density}_1 \times \frac{(\text{Distance}_1)^2}{(\text{Distance}_2)^2}$$

$$\text{Power Density}_2 = 10\,\frac{\text{mW}}{\text{m}^2} \times \frac{(3\lambda)^2}{(6\lambda)^2}$$

$$\text{Power Density}_2 = 10\,\frac{\text{mW}}{\text{m}^2} \times \frac{9\lambda^2}{36\lambda^2}$$

$$\text{Power Density}_2 = 10\,\frac{\text{mW}}{\text{m}^2} \times \frac{1}{4}$$

$$\text{Power Density}_2 = 2.5\,\frac{\text{mW}}{\text{m}^2}$$

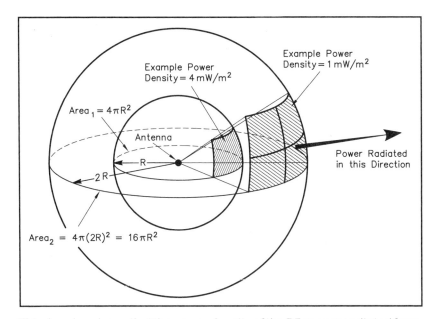

This drawing shows that the power density of the RF energy radiated from an antenna decreases as the square of the distance. This is because the area of a sphere increases as the square of the radius. ($A = 4\pi R^2$)

G0C05 If the free-space far-field power density of an antenna measures 9 milliwatts per square meter at a distance of 5 wavelengths, what will the field strength measure at a distance of 15 wavelengths?
   A. 3 milliwatts per square meter
   B. 1 milliwatt per square meter
   C. 0.9 milliwatt per square meter
   D. 0.09 milliwatt per square meter

B The power density decreases as the square of the distance from the antenna. When you are twice as far from the antenna, the electric field strength will be ¼ as strong. When you are 3 times farther away, the electric field strength will be $1/9$ as strong. We can write a simple equation based on this relationship to calculate the power density at some new distance when the power density is known at one distance.

$$\text{Power Density}_2 = 9 \, \frac{\text{mW}}{\text{m}^2} \times \frac{(5\lambda)^2}{(15\lambda)^2}$$

$$\text{Power Density}_2 = 9 \, \frac{\text{mW}}{\text{m}^2} \times \frac{25\lambda^2}{225\lambda^2}$$

$$\text{Power Density}_2 = 9 \, \frac{\text{mW}}{\text{m}^2} \times \frac{1}{9}$$

$$\text{Power Density}_2 = 1 \, \frac{\text{mW}}{\text{m}^2}$$

G0C06 What factors determine the location of the boundary between the near and far fields of an antenna?
   A. Wavelength of the signal and physical size of the antenna
   B. Antenna height and element material
   C. Boom length and element material
   D. Transmitter power and antenna gain

A The boundary between the near and far fields surrounding an antenna is a rather "fuzzy" area. Experts debate where the radiating near field ends and the radiating far field begins. In general, however, the wavelength of the energy and the physical size of the antenna are the most significant factors that help determine the boundary between the near and far radiating fields of an antenna.

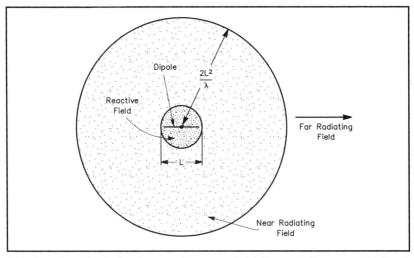

This drawing illustrates the reactive near field, the radiating near field and the far field around a half-wavelength dipole antenna.

G0C07 Which of the following steps might an amateur operator take to ensure compliance with the RF safety regulations?
   A. Post a copy of FCC Part 97 in the station
   B. Post a copy of OET Bulletin 65 in the station
   C. Nothing; amateur compliance is voluntary
   D. Perform a routine RF exposure evaluation

D Even if your station is exempt from the requirement, you may want to do a simple RF Radiation Exposure Evaluation. The results would demonstrate to yourself and possibly to your neighbors that your station is within the guidelines and is no cause for concern.

RF Safety Principles   211

G0C08 In the free-space far field, what is the relationship between the electric field (E field) and magnetic field (H field)?
   A. The electric field strength is equal to the square of the magnetic field strength
   B. The electric field strength is equal to the cube of the magnetic field strength
   C. The electric and magnetic field strength has a fixed impedance relationship of 377 ohms
   D. The electric field strength times the magnetic field strength equals 377 ohms

C If you have access to a laboratory-grade calibrated field-strength meter, several factors can upset the readings. Reflections from the ground and nearby conductors (power lines, other antennas, house wiring, etc.) can easily confuse field strength readings. You must also know the frequency response of the test equipment and probes, and use them only within the appropriate range. Even the orientation of the test probe with respect to the test antenna polarization is important. This calculation may prove useful to you as you analyze your station for compliance with the MPE limits. If you know the E or H field strength at some point in the far field then you can calculate the other value at that same point.

G0C09 What type of instrument can be used to accurately measure an RF field?
   A. A receiver with an S meter
   B. A calibrated field-strength meter with a calibrated antenna
   C. A betascope with a dummy antenna calibrated at 50 ohms
   D. An oscilloscope with a high-stability crystal marker generator

B You can use a calibrated field-strength meter and calibrated field-strength sensor (antenna) to accurately measure an RF field. Even if you have access to such an expensive laboratory-grade field-strength meter, several factors can upset the readings. Reflections from the ground and nearby conductors (power lines, other antennas, house wiring, etc.) can easily confuse field-strength readings. You must know the frequency response of the test equipment and probes, and use them only within the appropriate range. Even the orientation of the test probe with respect to the polarization of the antenna being tested is important.

G0C10 If your station complies with the RF safety rules and you reduce its power output from 500 to 40 watts, how would the RF safety rules apply to your operations?
- A. You would need to reevaluate your station for compliance with the RF safety rules because the power output changed
- B. You would need to reevaluate your station for compliance with the RF safety rules because the transmitting parameters changed
- C. You would not need to perform an RF safety evaluation, but your station would still need to be in compliance with the RF safety rules
- D. The RF safety rules would no longer apply to your station because it would be operating with less than 50 watts of power

C Any time you make changes to your station you should decide if you have done anything to increase the RF exposure to any area. Regardless of whether you need to do a new evaluation or not, your station must always comply with the RF safety rules.

G0C11 If your station complies with the RF safety rules and you reduce its power output from 1000 to 500 watts, how would the RF safety rules apply to your operations?
- A. You would need to reevaluate your station for compliance with the RF safety rules because the power output changed
- B. You would need to reevaluate your station for compliance with the RF safety rules because the transmitting parameters changed
- C. You would need to perform an RF safety evaluation to ensure your station would still be in compliance with the RF safety rules
- D. Since your station was in compliance with RF safety rules at a higher power output, you need to do nothing more

D Any time you make changes to your station you should decide if you have done anything to increase the RF exposure to any area. Regardless of whether you need to do a new evaluation or not, your station must always comply with the RF safety rules.

G0D Practical RF-safety applications

G0D01 Considering RF safety, what precaution should you take if you install an indoor transmitting antenna?
  A. Locate the antenna close to your operating position to minimize feed line losses
  B. Position the antenna along the edge of a wall where it meets the floor or ceiling to reduce parasitic radiation
  C. Locate the antenna as far away as possible from living spaces that will be occupied while you are operating
  D. Position the antenna parallel to electrical power wires to take advantage of parasitic effects

C   You should locate an antenna (whether it is indoors or outdoors) as far away as possible from living spaces that will be occupied while you are operating. Minimizing feed line radiation into populated areas is also a good practice.

G0D02 Considering RF safety, what precaution should you take whenever you make adjustments to the feed line of a directional antenna system?
  A. Be sure no one can activate the transmitter
  B. Disconnect the antenna-positioning mechanism
  C. Point the antenna away from the sun so it doesn't concentrate solar energy on you
  D. Be sure you and the antenna structure are properly grounded

A   Whenever you make adjustments around an antenna or feed line, you should take precautions to ensure that no one can turn on the transmitter.

G0D03 What is the best reason to place a protective fence around the base of a ground-mounted transmitting antenna?
  A. To reduce the possibility of persons being exposed to levels of RF in excess of the maximum permissible exposure (MPE) limits
  B. To reduce the possibility of animals damaging the antenna
  C. To reduce the possibility of persons vandalizing expensive equipment
  D. To improve the antenna's grounding system and thereby reduce the possibility of lightning damage

A   While answer A is the best and correct answer, B and C would also be secondary reasons for a protective fence. Only answer A relates to RF safety, however.

G0D04 What RF-safety precautions should you take before beginning repairs on an antenna?
  A. Be sure you and the antenna structure are grounded
  B. Be sure to turn off the transmitter and disconnect the feed-line
  C. Inform your neighbors so they are aware of your intentions
  D. Turn off the main power switch in your house

B One way to be sure that no one can activate the transmitter while you are working on it is to turn off the transmitter power supply and disconnect the antenna feed line.

G0D05 What precaution should be taken when installing a ground-mounted antenna?
  A. It should not be installed higher than you can reach
  B. It should not be installed in a wet area
  C. It should be painted so people or animals do not accidentally run into it
  D. It should be installed so no one can be exposed to RF radiation in excess of the maximum permissible exposure (MPE) limits

D No one should be near a transmitting antenna while it is in use. Install ground mounted transmitting antennas well away from living areas so that people cannot come so close that they receive more than the MPE limits.

G0D06 What precaution should you take before beginning repairs on a microwave feed horn or waveguide?
  A. Wear tight-fitting clothes and gloves to protect your body and hands from sharp edges
  B. Be sure the transmitter is turned off and the power source is disconnected
  C. Wait until the weather is dry and sunny
  D. Be sure propagation conditions are not favorable for troposphere ducting

B Feed horns and waveguides concentrate the transmitter output, and may result in hazardous levels of RF energy at the open end. In the UHF/SHF region, never look into the open end of an activated length of waveguide or microwave feed-horn antenna or point it toward anyone. Before working on a microwave feed horn or waveguide, turn off the transmitter and disconnect the feed line and power source.

G0D07 Why should directional high-gain antennas be mounted higher than nearby structures?
A. To eliminate inversion of the major and minor lobes
B. So they will not damage nearby structures with RF energy
C. So they will receive more sky waves and fewer ground waves
D. So they will not direct excessive amounts of RF energy toward people in nearby structures

D   All antennas should be mounted in a location that will ensure that no one can receive more than the MPE limits of RF radiation. This is especially important for directional, high-gain antennas. Such antennas should be mounted higher than nearby structures so the antenna won't direct excessive amounts of RF energy toward people in those buildings.

G0D08 For best RF safety, where should the ends and center of a dipole antenna be located?
A. Near or over moist ground so RF energy will be radiated away from the ground
B. As close to the transmitter as possible so RF energy will be concentrated near the transmitter
C. As far away as possible to minimize RF exposure to people near the antenna
D. Close to the ground so simple adjustments can be easily made without climbing a ladder

C   All antennas should be mounted in a location that will ensure that no one can receive more than the MPE limits of RF radiation.

G0D09 What should you do to reduce RF radiation exposure when operating at 1270 MHz?
A. Make sure that an RF leakage filter is installed at the antenna feed point
B. Keep the antenna away from your eyes when RF is applied
C. Make sure the standing wave ratio is low before you conduct a test
D. Never use a shielded horizontally polarized antenna

B   Your head and structures such as your eyes will absorb energy in the 1270 MHz range and higher frequencies more readily than in the HF or VHF range. To reduce RF radiation exposure when you are operating at these frequencies, keep the antenna away from your head and eyes when RF is applied.

GOD10 For best RF safety for driver and passengers, where should the antenna of a mobile VHF transceiver be mounted?
- A. On the right side of a metal rear bumper
- B. On the left side of a metal rear bumper
- C. In the center of a metal roof
- D. On the top-center of the rear window glass

C   The best place to mount the antenna for a mobile installation is in the center of a metal roof. The roof will provide an excellent shield to prevent the driver and passengers from being exposed to excessive amounts of RF energy. This is most practical for a VHF or UHF installation, but it may not be possible for an HF station because of the overall height of the vehicle and antenna. (Unless it is possible to measure the RF fields inside the vehicle, you should avoid transmitting with more than 25 W in a VHF mobile installation.)

GOD11 Considering RF safety, which of the following is the best reason to mount the antenna of a mobile VHF transceiver in the center of a metal roof?
- A. The roof will greatly shield the driver and passengers from RF radiation
- B. The antenna will be out of the driver's line of sight
- C. The center of a metal roof is the sturdiest mounting place for an antenna
- D. The wind resistance of the antenna will be centered between the wheels and not drag on one side or the other

A   The best place to mount the antenna for a mobile installation is in the center of a metal roof. The roof will provide an excellent shield to prevent the driver and passengers from being exposed to excessive amounts of RF energy. This is most practical for a VHF or UHF installation, but it may not be possible for an HF station because of the overall height of the vehicle and antenna. (Unless it is possible to measure the RF fields inside the vehicle, you should avoid transmitting with more than 25 W in a VHF mobile installation.)

G0E RF-safety solutions

G0E01 If you receive minor burns every time you touch your microphone while you are transmitting, which of the following statements is true?
    A. You need to use a low-impedance microphone
    B. You and others in your station may be exposed to more than the maximum permissible level of RF radiation
    C. You need to use a surge suppressor on your station transmitter
    D. All of these choices are correct

B   Commonly called "RF in the shack", minor burns or tingling are signs of unwanted RF in your station.

G0E02 If measurements indicate that individuals in your station are exposed to more than the maximum permissible level of radiation, which of the following corrective measures would be effective?
    A. Ensure proper grounding of the equipment
    B. Ensure that all equipment covers are tightly fastened
    C. Use the minimum amount of transmitting power necessary
    D. All of these choices are correct

D   If measurements indicate that people in your station may be exposed to more than the MPE limits of RF radiation, there are a number of steps you can take to reduce the exposure. Be sure that all equipment is properly grounded. Check all equipment covers to be sure they are securely fastened. Reducing transmitter power will also help reduce the exposure.

G0E03 If calculations show that you and your family may be receiving more than the maximum permissible RF radiation exposure from your 20-meter indoor dipole, which of the following steps might be appropriate?
    A. Use RTTY instead of CW or SSB voice emissions
    B. Move the antenna to a safe outdoor environment
    C. Use an antenna-matching network to reduce your transmitted SWR
    D. All of these choices are correct

B   Indoor antennas are often located close to living areas. This increases the chance for RF radiation exposure that is greater than the MPE limits. Moving your antenna to a safe outdoor location should be the first option you consider when you look for ways to reduce this excessive exposure. (Moving your indoor antenna to a safe outside environment may be difficult if the reason you have it indoors is that there is no safe outside environment.)

G0E04 Considering RF exposure, which of the following steps should you take when installing an antenna?
  A. Install the antenna as high and far away from populated areas as possible
  B. If the antenna is a gain antenna, point it away from populated areas
  C. Minimize feed line radiation into populated areas
  D. All of these choices are correct

D You should install all antennas as high and far away from populated areas as possible to minimize exposure to RF radiation. You also can point gain antennas away from populated areas. Minimizing feed line radiation into populated areas is also a good practice. All of these steps will help reduce the exposure of people to the RF radiation from your station.

G0E05 What might you do if an RF radiation evaluation shows that your neighbors may be receiving more than the maximum RF radiation exposure limit from your Yagi antenna when it is pointed at their house?
  A. Change from horizontal polarization to vertical polarization
  B. Change from horizontal polarization to circular polarization
  C. Use an antenna with a higher front to rear ratio
  D. Take precautions to ensure you can't point your antenna at their house

D A simple way to ensure that you cannot point your antenna toward a neighbor's house while your are transmitting is to make a few marks on your rotator control to remind you.

G0E06 What might you do if an RF radiation evaluation shows that your neighbors may be receiving more than the maximum RF radiation exposure limit from your quad antenna when it is pointed at their house?
  A. Reduce your transmitter power to a level that reduces their exposure to a value below the maximum permissible exposure (MPE) limit
  B. Change from horizontal polarization to vertical polarization
  C. Use an antenna with a higher front to side ratio
  D. Use an antenna with a sharper radiation lobe

A In addition to ensuring that you can't point your gain antenna at your neighbor's house, you also can reduce their exposure to RF radiation from your station by reducing your transmitter power. If you reduce your transmitter power to a level that reduces their exposure to a value below the MPE limits, you may safely be able to continue operating even with the antenna pointed in that direction.

G0E07 Why does a dummy antenna provide an RF safe environment for transmitter adjusting?
- A. The dummy antenna carries the RF energy far away from the station before releasing it
- B. The RF energy is contained in a halo around the outside of the dummy antenna
- C. The RF energy is not radiated from a dummy antenna, but is converted to heat
- D. The dummy antenna provides a perfect match to the antenna feed impedance

C A dummy antenna is best when you have to make adjustments to your transmitter while it is operating. The RF energy is converted to heat rather than being radiated. It also allows you to make the adjustments without putting a signal on the air.

G0E08 From an RF radiation exposure point of view, which of the following materials would be the best to use for your homemade transmatch enclosure?
- A. Aluminum
- B. Bakelite
- C. Transparent acrylic plastic
- D. Any nonconductive material

A A commercial transmatch network would come with a metal case enclosing it. It is important to remember that if you are building your own, it needs to be securely enclosed in aluminum or other metal conductive material.

G0E09 From an RF radiation exposure point of view, what is the advantage to using a high-gain, narrow-beamwidth antenna for your VHF station?
- A. High-gain antennas absorb stray radiation
- B. The RF radiation can be focused in a direction away from populated areas
- C. Narrow-beamwidth antennas eliminate exposure in areas directly under the antenna
- D. All of these choices are correct

B The whole idea is to keep people's exposure to RF radiation below the limits prescribed in the regulations. The direction your antenna is pointing and its main beam location, its height above the ground, its gain and beamwidth, and your transmitter power all must be considered.

GOE10 From an RF radiation exposure point of view, what is the disadvantage in using a high-gain, narrow-beamwidth antenna for your VHF station?
- A. High-gain antennas must be fed with coaxial cable feed-line, which radiates stray RF energy
- B. The RF radiation can be better focused in a direction away from populated areas
- C. Individuals in the main beam of the radiation pattern will receive a greater exposure than when a low-gain antenna is used
- D. All of these choices are correct

C While a high-gain, narrow beamwidth antenna allows you to concentrate the radiated RF energy from your station in a certain direction, this also creates a disadvantage to this type of antenna. Anyone who happens to be in the main beam of the radiation from that antenna will receive a higher RF exposure than they would with a lower gain antenna.

GOE11 If your station is located in a residential area, which of the following would best help you reduce the RF exposure to your neighbors from your amateur station?
- A. Use RTTY instead of CW or SSB voice emissions
- B. Install your antenna as high as possible to maximize the distance to nearby people
- C. Use top-quality coaxial cable to reduce RF losses in the feedline
- D. Use an antenna matching network to reduce your transmitted SWR

B The whole idea is to keep people's exposure to RF radiation below the limits prescribed in the regulations. The direction your antenna is pointing and its main beam location, its height above the ground, its gain and beamwidth, and your transmitter power all must be considered.

About the ARRL

The seed for Amateur Radio was planted in the 1890s, when Guglielmo Marconi began his experiments in wireless telegraphy. Soon he was joined by dozens, then hundreds, of others who were enthusiastic about sending and receiving messages through the air—some with a commercial interest, but others solely out of a love for this new communications medium. The United States government began licensing Amateur Radio operators in 1912.

By 1914, there were thousands of Amateur Radio operators—hams—in the United States. Hiram Percy Maxim, a leading Hartford, Connecticut, inventor and industrialist saw the need for an organization to band together this fledgling group of radio experimenters. In May 1914 he founded the American Radio Relay League (ARRL) to meet that need.

Today ARRL, with approximately 170,000 members, is the largest organization of radio amateurs in the United States. The ARRL is a not-for-profit organization that:

- promotes interest in Amateur Radio communications and experimentation
- represents US radio amateurs in legislative matters, and
- maintains fraternalism and a high standard of conduct among Amateur Radio operators.

At ARRL headquarters in the Hartford suburb of Newington, the staff helps serve the needs of members. ARRL is also International Secretariat for the International Amateur Radio Union, which is made up of similar societies in 150 countries around the world.

ARRL publishes the monthly journal *QST*, as well as newsletters and many publications covering all aspects of Amateur Radio. Its headquarters station, W1AW, transmits bulletins of interest to radio amateurs and Morse code practice sessions. The ARRL also coordinates an extensive field organization, which includes volunteers who provide technical information and other support services for radio amateurs as well as communications for public-service activities. In addition, ARRL represents US amateurs with the Federal Communications Commission and other government agencies in the US and abroad.

Membership in ARRL means much more than receiving *QST* each month. In addition to the services already described, ARRL offers membership services on a personal level, such as the ARRL Volunteer Examiner Coordinator Program and a QSL bureau.

Full ARRL membership (available only to licensed radio amateurs) gives you a voice in how the affairs of the organization are governed. ARRL policy is set by a Board of Directors (one from each of 15 Divisions). Each year, one-third of the ARRL Board of Directors stands for election by the full members they represent. The day-to-day operation of ARRL HQ is managed by an Executive Vice President and his staff.

No matter what aspect of Amateur Radio attracts you, ARRL membership is relevant and important. There would be no Amateur Radio as we know it today were it not for the ARRL. We would be happy to welcome you as a member! (An Amateur Radio license is not required for Associate Membership.) For more information about ARRL and answers to any questions you may have about Amateur Radio, write or call:

ARRL—The national association for Amateur Radio
225 Main Street
Newington CT 06111-1494
Voice: 860-594-0200
Fax: 860-594-0259
E-mail: **hq@arrl.org**
Internet: **www.arrl.org/**

Prospective new amateurs call (toll-free):
800-32-NEW HAM (800-326-3942)
You can also contact us via e-mail at **newham@arrl.org**
or check out *ARRLWeb* at **http://www.arrl.org/**

Join ARRL and experience the BEST of Ham Radio!

I want to join ARRL. My membership includes:

- **QST**—12 monthly issues of **Ham Radio's #1 Magazine**
- **Technical Information Service, Operating Awards**, and support services for volunteers
- **ARRL Web features for Members Only** and email forwarding service ("your-callsign"@arrl.net)
- **"All Risk" Ham Radio Equipment Insurance**-coverage available
- **Representation of the Amateur Radio Service** before national and international government

☐ New member   ☐ Previous member   ☐ Renewal

Call Sign (if any)   Class of License   Date of Birth

Name

Address

City, State, ZIP

Dues are $39 per year in the US. You do not need an Amateur Radio license to join. Individuals age 65 or over, residing in the US, upon submitting date of birth, may request the dues rate of $34. Immediate relatives of a member who receives QST, and reside at the same address may request family membership at $8 per year. Blind individuals may join without QST for $8 per year. If you are 21 or younger and a licensed amateur, a special rate may apply. Write or call ARRL for details.

DUES ARE SUBJECT TO CHANGE WITHOUT NOTICE.

Payment to ARRL enclosed ☐

Charge to MC, VISA, AMEX, Discover No. _____

Expiration Date _____

Cardholder Name _____

Cardholder Signature _____

If you do not wish your name and address made available for non-ARRL related mailings, please check this box. ☐

ARRL *The national association for* **AMATEUR RADIO**

225 MAIN STREET NEWINGTON, CONNECTICUT 06111 USA
Call toll free to join: (888) 277-5289
Join on the Web: www.arrl.org/join.html

GQA 02

Please use this form to give us your comments on this book and what you'd like to see in future editions, or e-mail us at **pubsfdbk@arrl.org** (publications feedback). If you use e-mail, please include your name, call, e-mail address and the book title, edition and printing in the body of your message. Also indicate whether or not you are an ARRL member.

Please check the box that best answers these questions:
How well did this book prepare you for your exam?
☐ Very Well ☐ Fairly Well ☐ Not Very Well
Did you pass? ☐ Yes ☐ No
Where did you purchase this book?
☐ From ARRL directly ☐ From an ARRL dealer

Is there a dealer who carries ARRL publications within:
☐ 5 miles ☐ 15 miles ☐ 30 miles of your location? ☐ Not sure.

If licensed, what is your license class? _____

Name _____ ARRL member? ☐ Yes ☐ No
_____ Call Sign _____
Address _____
City, State/Province, ZIP/Postal Code _____
Daytime Phone (   ) _____ Age _____ E-mail _____
If licensed, how long? _____

For ARRL use only	G Q&A
Edition	1 2 3 4 5 6 7 8 9 10 11
Printing	1 2 3 4 5 6 7 8 9 10 11

Other hobbies _____
_____
Occupation _____

From _____
_____
_____

Please affix postage. Post Office will not deliver without postage.

EDITOR, THE ARRL'S GENERAL Q&A
ARRL—THE NATIONAL ASSOCIATION FOR AMATEUR RADIO
225 MAIN STREET
NEWINGTON CT 06111-1494

— — — — — — — — — please fold and tape — — — — — — — — —